Hello, happy to see you here.

I'm BB, aka Bead Baby.

Abacus is fun, if you know All About it.

"How to Abacus" has books to help you.

Book 1: All About Abacus

Book 2: All About Formulas

Book 3: All About Times and Division

Book 4: All About Mental Math.

I will definitely be with you through all the lessons.

Let's go!

AF811572

To those who are interested in Abacus. ~ Sheena

Sheena, an Abacus teacher, also a media veteran,

is passionate to share the benefits of Abacus to everyone.

For anyone willing to engage in Abacus Learning. ~ Hanlong

Hanlong started to learn Abacus at sixth grade.

She teaches Abacus to share her Abacus experiences with children

Text and pictures copyright © 2018 by Sheena Chin and Hanlong Fan

All rights reserved.

No parts of this book may be used, reproduced, scanned or transmitted in any form or by any means, electronic or mechanical, including photocopying or recording, without written permission from the publisher.

For information address PinGrow Media, 5205 Prospect Road # 135 - 368, San Jose California 95129 or contact@pingrow.com

ISBN-13: 978-1-949622-02-7

Visit www.pingrow.com

How to Abacus 2

All About Formulas

Table of Content:

Chapter 1: Three Groups of Friends ... 5

Chapter 2: Big Friends 6
Who is my Big Friend?

Chapter 3: Little Friends 8
Little Friends Practice

Chapter 4: Mixed Friends 10
When to use Mixed Friends

Chapter 5: Big Friends Formulas 12
1&9, 2&8, 3&7, 4&6
Exercises

Chapter 6: Little Friends Formulas 44
1&4, 2&3
Exercises

Chapter 7: Mixed Friends Formulas .. 58
6, 7, 8 and 9
Exercises

What is Formula? Isn't it for babies?
No, it's the secret for Abacus.
Really?
Sure, let's start digging out the secret.

Chapter 1
3 Groups of Friends:

In order to get Abacus secrets,
you need to be a good friend with the three groups of friends: **Big Friends, Little Friends and Mixed Friends**.

Formulas for the three groups of Friends will be introduced in chapter 5, 6 and 7 respectively and need to be **memorized**.

Big Friends:
Two numbers add together to be 10.

Little Friends:
Two numbers add together to be 5.

Mixed Friends:
Big Friends + Little Friends

Chapter 2

Big Friends

When two numbers add together to be 10, these two numbers are Big Friends for each other. For instance 1 + 9 = 10, so 9's Big Friend is 1 and 1's Big Friend is 9; 2 + 8 = 10, so 8 and 2 are mutual Big Friends. Therefore, 3 & 7 and 4 & 6 are also Big Friends respectively. 5 + 5 = 10, so 5's Big Friend is its twin sibling, 5.

Add to be 10	Mutual Big Friends
1 + 9 = 10	1, 9
2 + 8 = 10	2, 8
3 + 7 = 10	3, 7
4 + 6 = 10	4, 6
5 + 5 = 10	5, 5

Who is my Big Friend?

Draw a line to connect Big Friends.

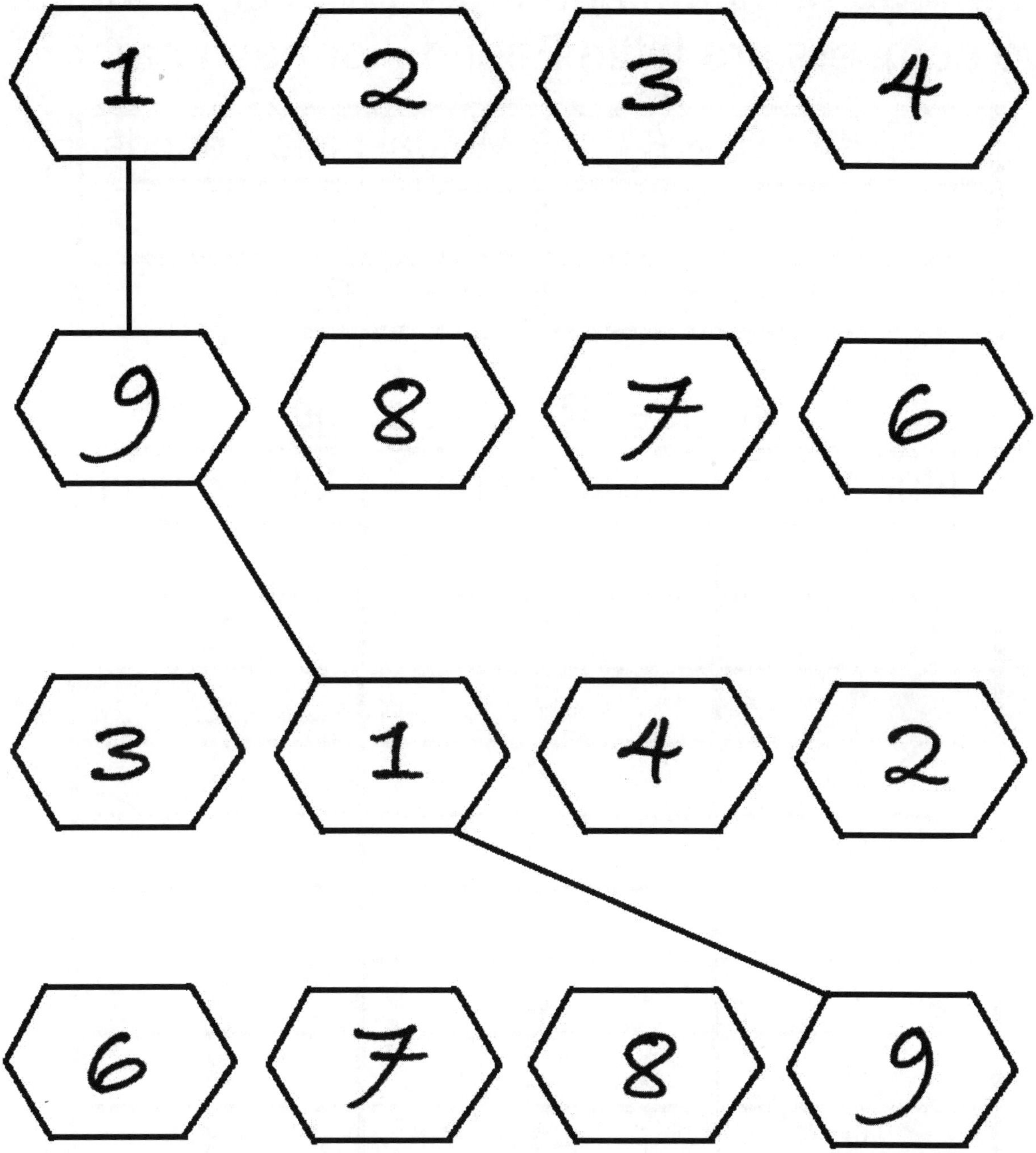

Chapter 3
Little Friends

When two numbers add together to be 5, these two numbers are Little Friends for each other.[1]

Add to be 5	Mutual Little Friends
1 + 4 = 5	1, 4
2 + 3 = 5	2, 3

Number Relationships

Numbers	Big Friend	Little Friend
1	9	4
2	8	3
3	7	2
4	6	1
5	5	X
6	4	X
7	3	X
8	2	X
9	1	X

[1] It involves one movement method. Please refer to How to Abacus 1, All About Abacus p. 30

Little Friends Practice

Circle Little Friends

2 8 4 7 6

3 5 1 9

Connect Little Friends

1

2 4

3

Chapter 4
Mixed Friends

Mixed Friends is a combination of Big Friends and Little Friends. Mixed Friends **only** involve with bigger numbers such as 6, 7, 8 and 9.

Mixed Friends Formulas emerge when Big Friends Formulas cannot solve problems alone and need the help of Little Friends Formulas. Mixed Friends Formulas usually involves with 5 and happens when there are not enough beads to be used on ones place while applying Big Friends Formulas and when adding more beads back on ones place after applying Big Friends Formulas.

Illustrations on the next page demonstrate two conditions for Big Friends Formulas to combine with Little Friends Formulas to generate Mixed Friends Formulas. The concepts of the following examples also apply to + 7, - 7, + 8, - 8, + 9 and -9.

When to use Mixed Friends

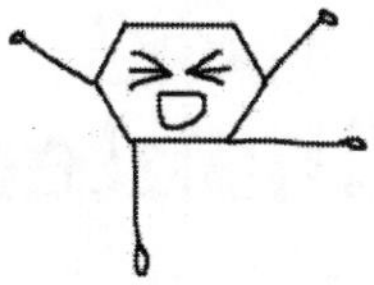

Not enough beads to be used on ones place

Problem: 5 + 6 = ?

5 + 6 = - 4 + 10

does not have 4 bottom beads to be subtracted by Big Friends
+ 6 = - 4 + 10

combine with Little Friends
- 4 = - 5 + 1

Mixed Friends
+ 6 = - 5 + 1 + 10

Problem: 5 **+ 6** = 5 **- 5 + 1 + 10** = 11

Add more beads back on ones place

Problem: 14 - 6 = ?

14 - 6 = - 10 + 4

4 bottom beads on ones place plus 4 ➡ (10 - 6)

combine with Little Friends
+ 4 = + 5 -1

Mixed Friends
- 6 = - 10 + 5 - 1

Problem: 14 **- 6** = 14 **- 10 + 5 - 1** = 8

Chapter 5: Big Friends Formulas

Formulas for Big Friends 1 & 9

+ 1 = - 9 + 10

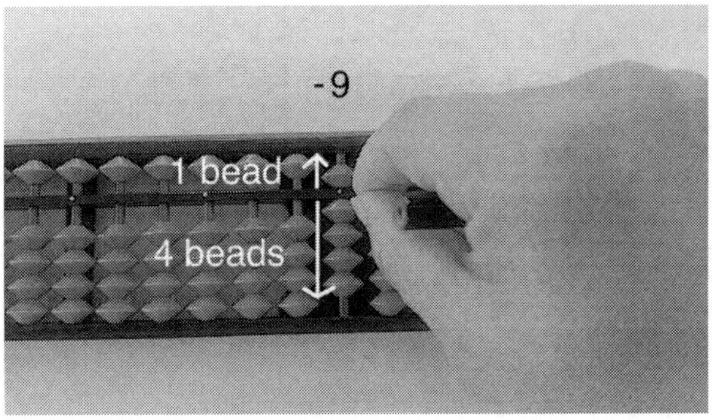

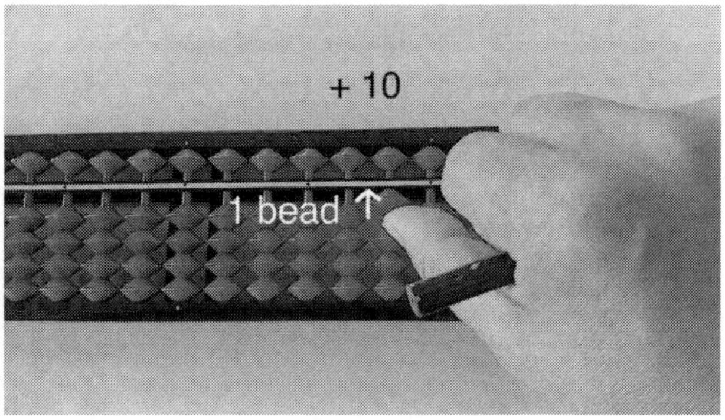

Formula **+ 1 = - 9 + 10** Exercises

1	2	3	4	5	6	7	8
9 1	19 1	29 1	39 1	59 1	69 1	79 1	89 1

1	2	3	4	5	6	7	8	9	10
2 7 1	3 6 1	5 4 1	8 1 1	7 2 1	6 3 1	4 5 1	1 8 1	12 7 1	13 6 1

Formulas for Big Friends 1 & 9

- 1 = - 10 + 9

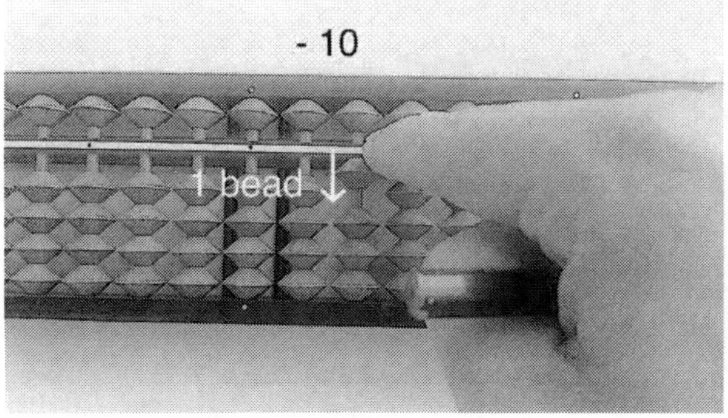

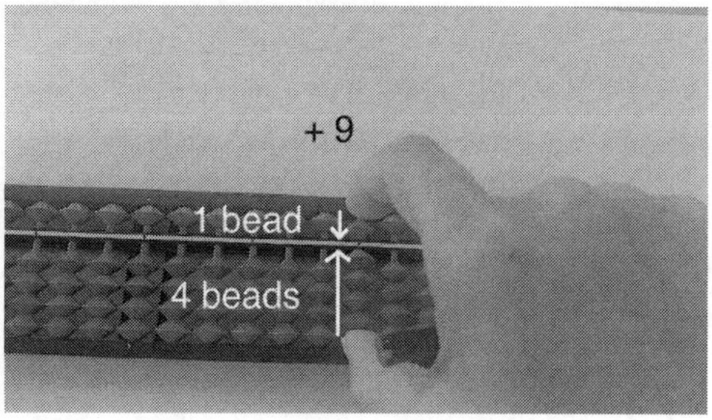

Formula - **1** = - **10** + **9** Exercises

1	2	3	4	5	6	7	8
10 -1	20 -1	30 -1	40 -1	60 -1	70 -1	80 -1	90 -1

1	2	3	4	5	6	7	8	9	10
11 -1 -1	13 -3 -1	12 -2 -1	19 -9 -1	21 -1 -1	14 -4 -1	18 -8 -1	16 -6 -1	15 -5 -1	17 -7 -1

Formulas for Big Friends 1 & 9

+ 9 = - 1 + 10

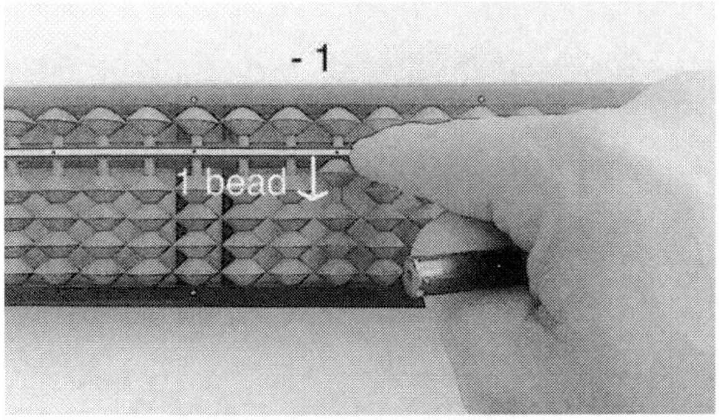

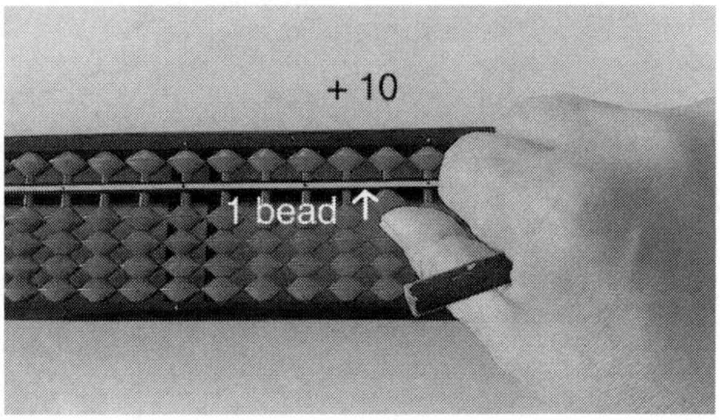

Formula **+ 9 = - 1 + 10** Exercises

1	2	3	4	5	6	7	8
1 9	2 9	3 9	4 9	6 9	7 9	8 9	9 9

1	2	3	4	5	6	7	8	9	10
2 1 9	3 -2 9	4 -2 9	3 -1 9	8 -2 9	5 2 9	1 1 9	4 -3 9	6 1 9	7 1 9

Formulas for Big Friends 1 & 9

- 9 = - 10 + 1

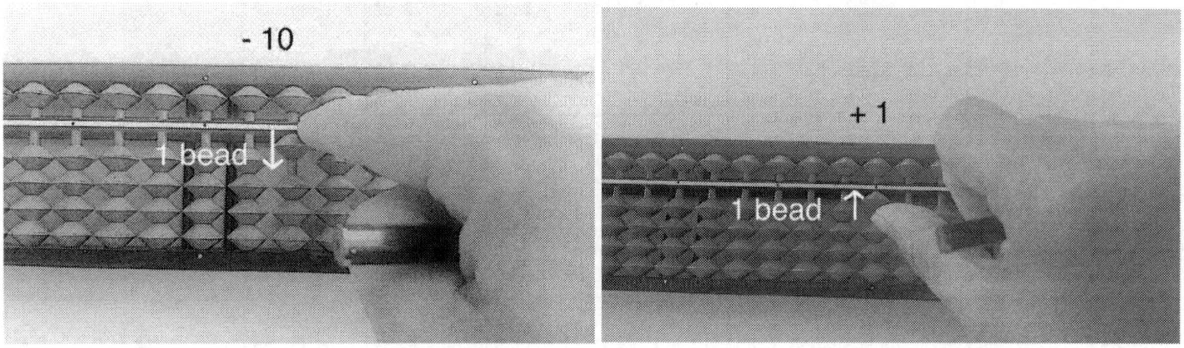

Formula **- 9 = - 10 + 1** Exercises

1	2	3	4	5	6	7	8
10 -9	11 -9	12 -9	13 -9	16 -9	17 -9	18 -9	20 -9

1	2	3	4	5	6	7	8	9	10
10 1 -9	11 -1 -9	12 -1 -9	13 -1 -9	14 -2 -9	16 -1 -9	17 -1 -9	18 -3 -9	12 1 -9	11 1 -9

Exercises: Big Friends 1 & 9

1	2	3	4	5	6	7	8	9	10
5	7	8	9	8	5	1	9	7	13
1	-2	-3	-4	-2	1	5	1	-1	-2
9	1	11	2	9	12	10	3	9	-1
2	9	-9	9	4	-9	-9	-9	-9	-9

11	12	13	14	15	16	17	18	19	20
6	9	8	16	8	5	3	9	4	8
11	1	-3	2	-2	2	5	-3	-1	-2
-7	6	2	-8	3	2	1	12	10	9
-9	-9	9	-1	1	1	9	-9	-9	3

21	22	23	24	25	26	27	28	29	30
15	6	12	5	7	1	2	6	9	10
1	3	1	2	-2	9	2	9	-1	-9
-6	9	-3	9	3	9	9	-5	9	2
-1	1	-1	2	9	1	1	-1	2	9

Exercises: Big Friends 1 & 9

1	2	3	4	5	6	7	8	9	10
1	3	6	5	9	4	8	2	11	7
5	6	2	3	1	9	1	5	-1	-2
9	-2	10	-2	6	-3	9	11	-9	11
-5	9	-8	9	3	-1	-8	-8	6	-6
-1	9	-1	4	-4	9	-9	-9	9	-9

11	12	13	14	15	16	17	18	19	20
11	10	2	5	3	4	7	1	6	9
6	-1	9	2	1	9	2	5	-1	1
9	9	-1	9	9	-3	11	9	12	3
-6	-8	-1	-6	-3	-9	-9	2	-7	9
-1	-1	9	-1	-9	1	3	9	-9	1

21	22	23	24	25	26	27	28	29	30
1	2	4	7	5	1	3	8	6	13
9	1	5	2	3	9	5	1	-1	-9
2	9	1	1	9	6	1	1	2	5
9	-2	-9	-1	-7	9	1	-9	9	1
-1	-1	-1	9	-1	1	1	-1	2	9

Formulas for Big Friends 2 & 8

+ 2 = - 8 + 10

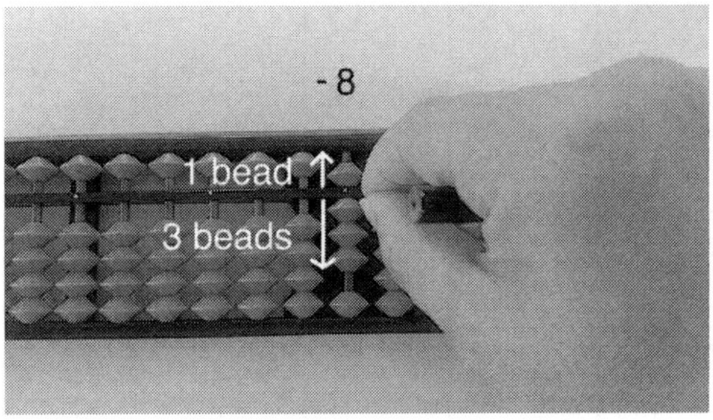

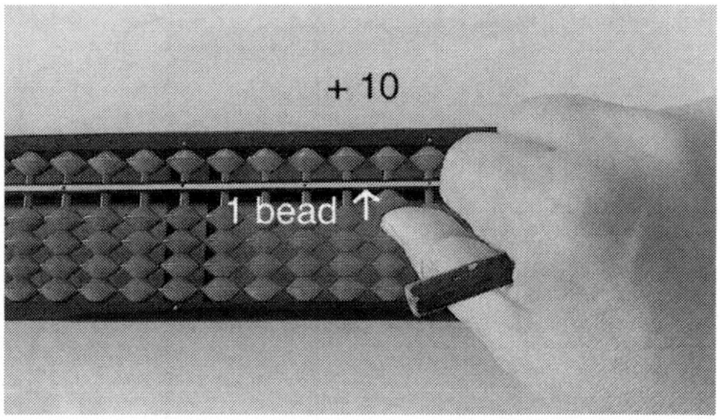

Formula **+ 2 = - 8 + 10** Exercises

1	2	3	4	5	6	7	8
8	18	28	38	9	19	29	39
2	2	2	2	2	2	2	2

1	2	3	4	5	6	7	8	9	10
2	3	7	6	5	2	3	11	11	12
6	5	1	2	4	7	6	7	8	6
2	2	2	2	2	2	2	2	2	2

Formulas for Big Friends 2 & 8

- 2 = - 10 + 8

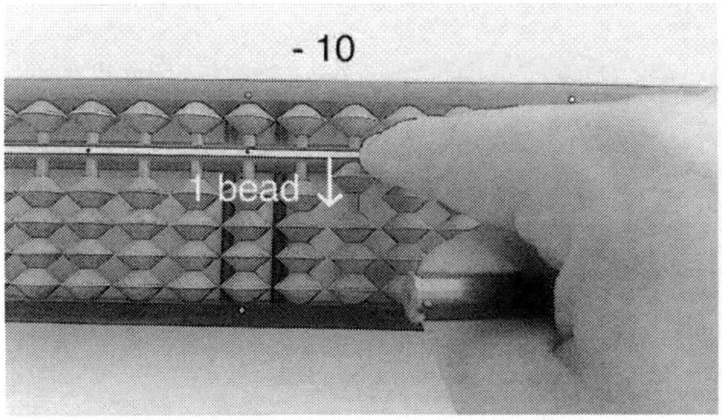

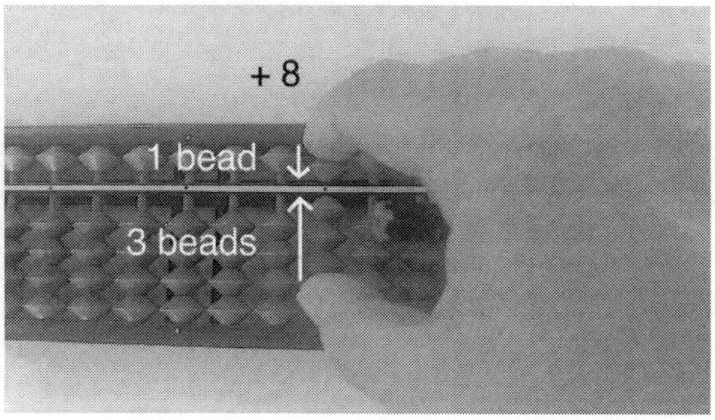

Formula - **2 = - 10 + 8** Exercises

1	2	3	4	5	6	7	8
10	20	30	40	11	21	31	41
-2	-2	-2	-2	-2	-2	-2	-2

1	2	3	4	5	6	7	8	9	10
10	11	12	13	14	16	17	18	19	20
1	-1	-1	-2	-3	-5	-6	-8	-9	1
-2	-2	-2	-2	-2	-2	-2	-2	-2	-2

Formulas for Big Friends 2 & 8

+ 8 = - 2 + 10

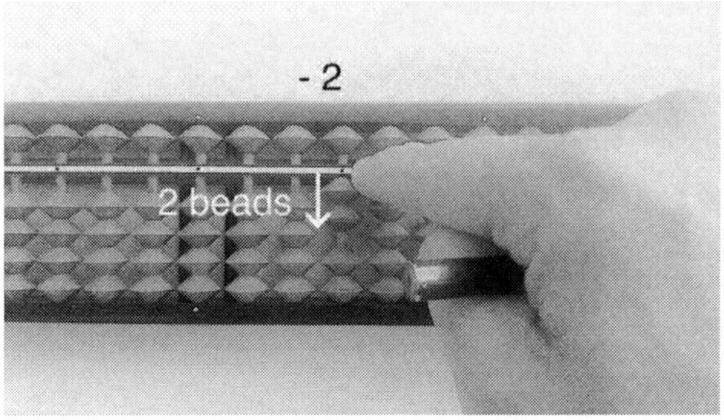

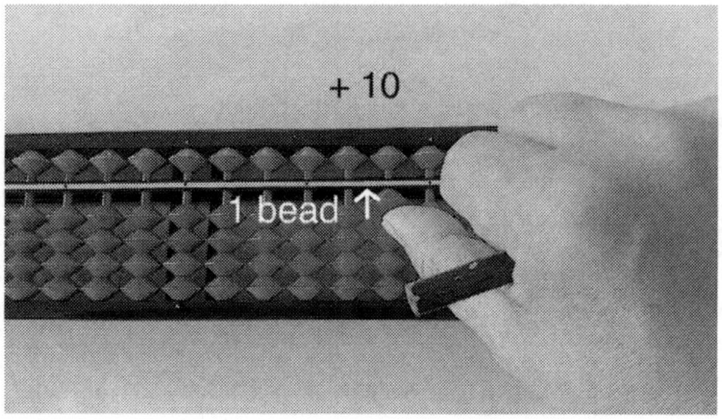

Formula **+ 8 = - 2 + 10** Exercises

1	2	3	4	5	6	7	8
2	3	4	7	8	9	12	17
8	8	8	8	8	8	8	8

1	2	3	4	5	6	7	8	9	10
2	3	4	3	8	5	1	4	12	11
1	1	-2	-1	1	2	1	-1	1	1
8	8	8	8	8	8	8	8	8	8

Formulas for Big Friends 2 & 8

- 8 = - 10 + 2

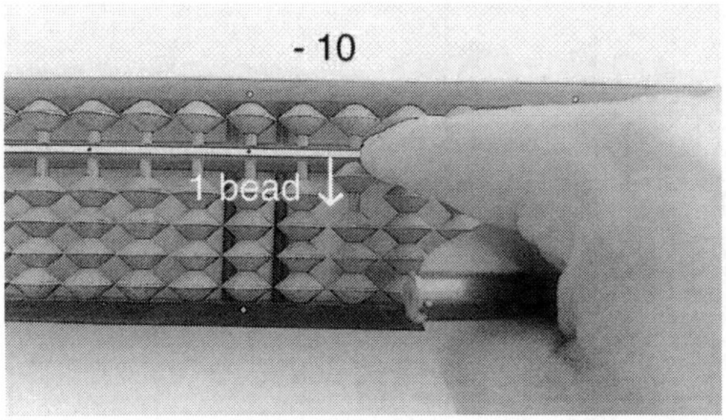

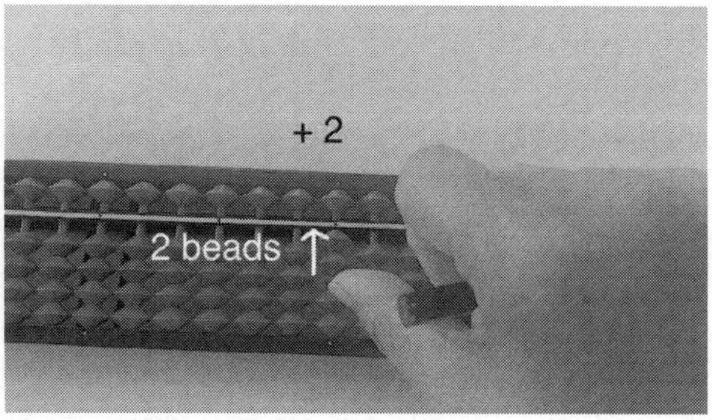

Formula - **8** = - **10** + **2** Exercises

1	2	3	4	5	6	7	8
10 -8	11 -8	12 -8	17 -8	20 -8	21 -8	22 -8	27 -8

1	2	3	4	5	6	7	8	9	10
10 2 -8	11 -1 -8	12 -1 -8	13 -1 -8	14 -2 -8	16 -1 -8	17 -1 -8	18 -3 -8	12 -1 -8	11 1 -8

Exercises: Big Friends 2 & 8

1	2	3	4	5	6	7	8	9	10
2	10	3	11	5	8	6	12	7	1
8	-2	1	-8	2	2	1	-8	2	8
8	1	8	1	8	1	8	5	2	8
2	8	-8	8	3	-8	1	8	8	8

11	12	13	14	15	16	17	18	19	20
9	1	11	4	2	10	3	5	8	7
8	3	-2	8	1	-8	8	4	1	8
1	8	8	6	8	6	6	2	2	-5
-3	5	1	-3	5	2	-8	-1	-1	-2

21	22	23	24	25	26	27	28	29	30
6	3	5	4	1	8	2	7	12	9
2	8	3	8	7	8	8	1	5	-2
8	-2	8	-2	8	-6	7	8	-8	8
-1	-1	-1	-8	-6	-2	-8	-6	2	3

Exercises: Big Friends 2 & 8

1	2	3	4	5	6	7	8	9	10
2	1	3	5	9	4	6	8	10	7
1	2	8	3	-6	-2	3	2	-2	-5
8	8	-1	2	8	8	2	6	8	8
-1	-2	-2	-8	-2	1	-8	2	3	8
-2	-3	8	6	-2	-2	1	2	2	2

11	12	13	14	15	16	17	18	19	20
2	9	3	4	7	1	6	5	8	6
8	-7	5	5	8	1	11	3	-6	-5
7	8	-6	2	-5	8	-8	8	8	10
-6	-2	8	-8	1	-2	2	-6	-2	-2
-2	1	-2	1	-2	-6	5	-2	-1	8

21	22	23	24	25	26	27	28	29	30
9	1	5	7	2	8	6	3	13	4
8	8	2	1	2	8	3	1	-2	-3
8	2	8	2	8	-6	8	8	-8	10
-5	-8	3	-8	6	-2	1	6	6	-2
-2	1	8	2	-3	1	2	2	8	-3

Exercises: Big Friends 1&9, 2&8

1	2	3	4	5	6	7	8	9	10
9	2	5	3	6	1	4	7	13	8
1	9	2	8	11	9	8	9	6	2
8	-8	9	-9	-9	-2	-9	-8	1	-1
2	5	-8	2	2	8	5	2	-8	8

11	12	13	14	15	16	17	18	19	20
1	8	2	5	7	3	9	4	6	2
2	-6	8	1	8	9	-7	9	-5	11
9	8	-9	9	-9	-8	8	8	9	8
-8	-9	6	-8	2	5	-1	-2	-8	-9

21	22	23	24	25	26	27	28	29	30
6	9	15	2	4	7	3	1	8	5
9	9	-8	1	-2	-6	8	1	2	2
-8	8	9	8	8	9	-9	8	-9	8
-5	-6	-8	-9	-1	-2	8	-1	6	-9

Exercises: Big Friends 1&9, 2&8

1	2	3	4	5	6	7	8	9	10
3	2	6	5	9	1	4	7	8	13
9	8	2	2	9	9	9	9	-1	-9
8	-9	8	9	2	7	8	-8	8	8
-2	6	9	-8	-8	8	-1	2	-9	-2
9	8	-5	2	1	-9	-2	-9	3	-2

11	12	13	14	15	16	17	18	19	20
9	10	4	7	2	8	1	3	5	6
1	-2	8	8	1	8	8	5	1	3
-2	9	8	-9	9	-9	2	2	9	8
8	-8	-9	12	-8	8	-9	-8	-8	-9
-6	1	-9	-9	-2	-9	2	9	2	8

21	22	23	24	25	26	27	28	29	30
12	3	1	9	6	4	7	2	5	8
8	5	1	8	2	-1	-5	9	3	2
-9	9	9	-7	8	8	8	-8	9	-1
-9	-8	-8	-1	-9	-9	-1	5	-8	2
1	1	9	2	8	8	9	9	1	-1

Formulas for Big Friends 3 & 7

+ 3 = - 7 + 10

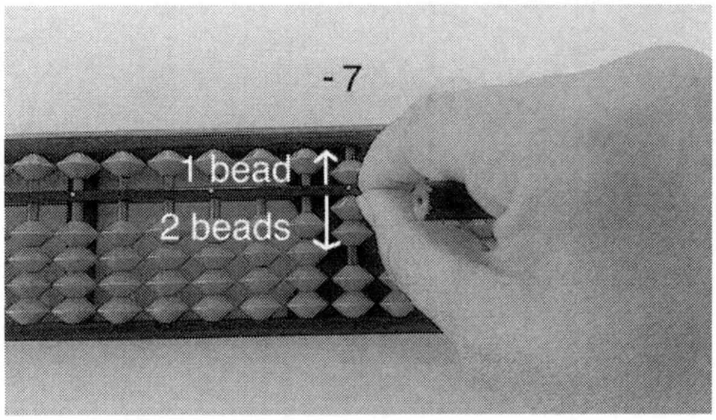

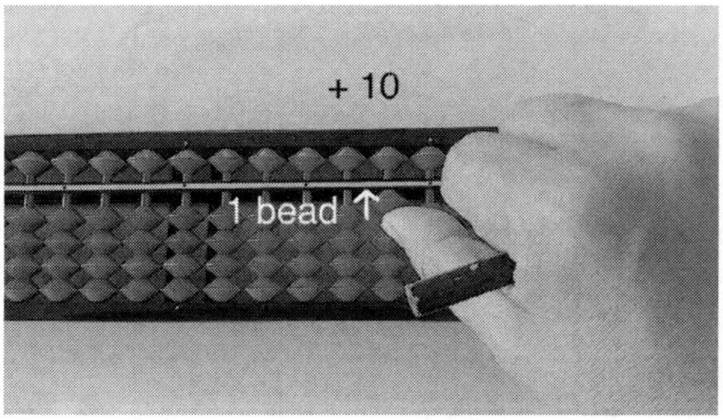

Formula **+ 3 = - 7 + 10** Exercises

1	2	3	4	5	6	7	8
7	8	9	17	18	19	27	28
3	3	3	3	3	3	3	3

1	2	3	4	5	6	7	8	9	10
2	3	4	1	2	5	3	8	7	2
5	5	5	6	6	2	6	1	1	7
3	3	3	3	3	3	3	3	3	3

Formulas for Big Friends 3 & 7

- 3 = - 10 + 7

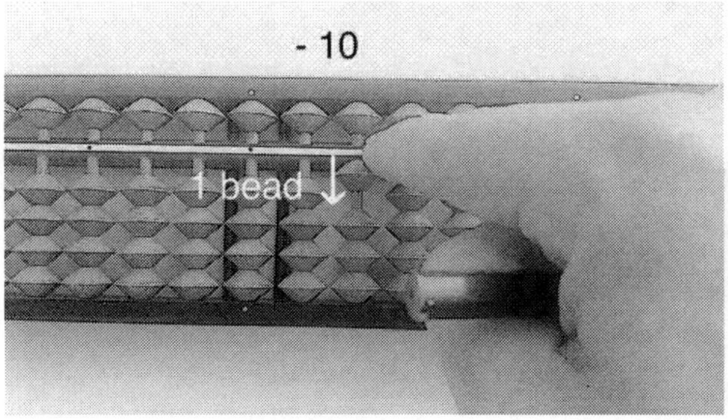

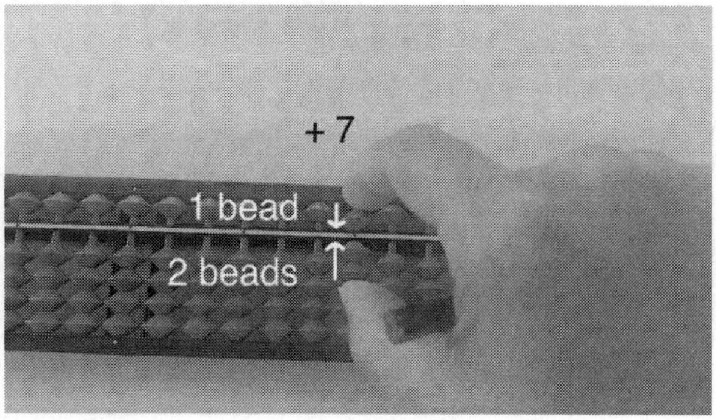

Formula **- 3 = - 10 + 7** Exercises

1	2	3	4	5	6	7	8
10 -3	11 -3	12 -3	20 -3	21 -3	22 -3	30 -3	31 -3

1	2	3	4	5	6	7	8	9	10
10 2 -3	11 -1 -3	12 -1 -3	13 -1 -3	14 -2 -3	16 -5 -3	17 -5 -3	18 -6 -3	12 -2 -3	11 1 -3

Formulas for Big Friends 3 & 7

+ 7 = - 3 + 10

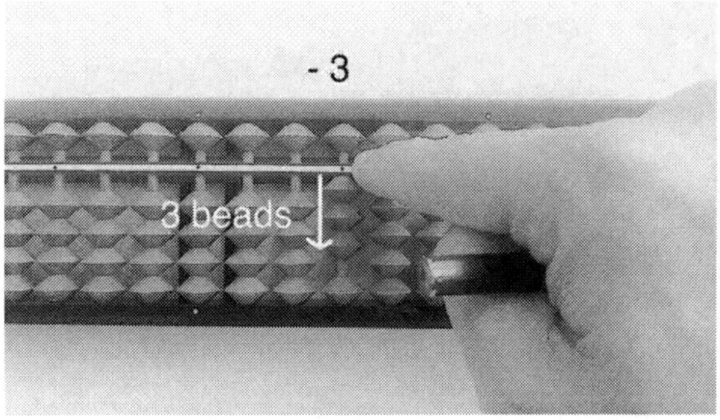

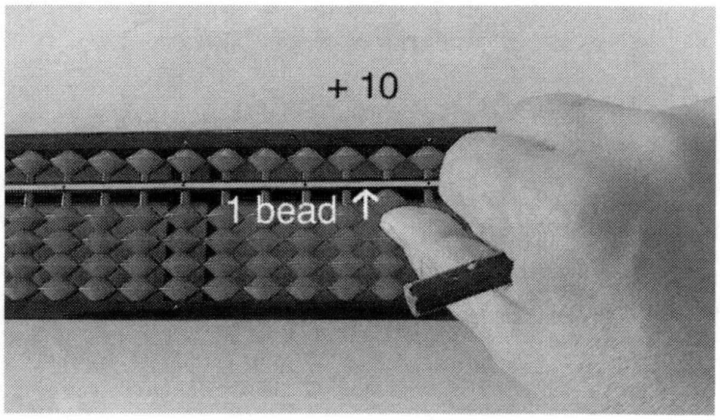

Formula **+ 7 = - 3 + 10** Exercises

1	2	3	4	5	6	7	8
3 7	8 7	9 7	13 7	18 7	19 7	23 7	28 7

1	2	3	4	5	6	7	8	9	10
3 5 7	3 6 7	4 5 7	1 2 7	3 1 7	2 2 7	1 7 7	2 6 7	2 7 7	8 1 7

Formulas for Big Friends 3 & 7

- 7 = - 10 + 3

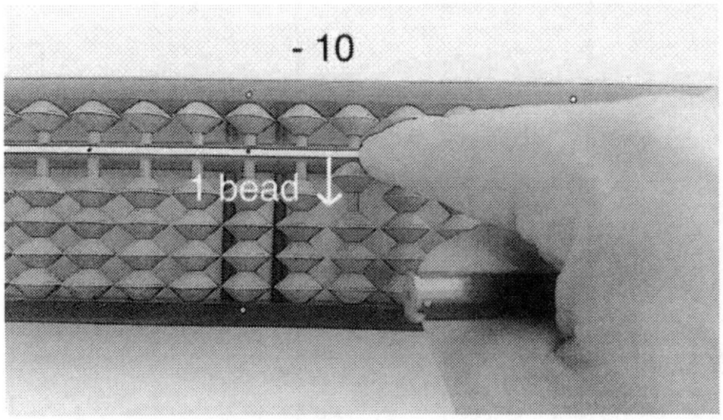

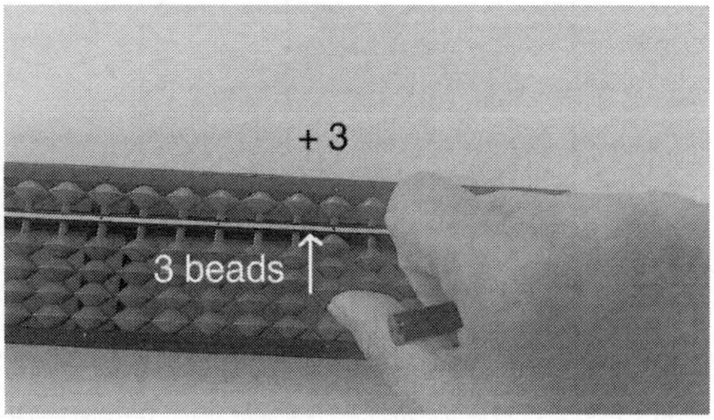

Formula **- 7 = - 10 + 3** Exercises

1	2	3	4	5	6	7	8
10 -7	11 -7	15 -7	16 -7	20 -7	21 -7	25 -7	26 -7

1	2	3	4	5	6	7	8	9	10
10 1 -7	11 -1 -7	12 -1 -7	13 -2 -7	14 -3 -7	16 -5 -7	17 -7 -7	16 -1 -7	17 -1 -7	18 -3 -7

Exercises: Big Friends 3 & 7

1	2	3	4	5	6	7	8	9	10
7	4	1	5	6	3	8	2	9	3
3	7	2	2	2	5	3	1	3	7
3	6	7	3	3	3	-7	7	1	5
7	3	-3	-7	-7	-7	5	-3	7	-7

11	12	13	14	15	16	17	18	19	20
2	8	6	7	4	3	1	9	5	10
2	-1	1	2	5	7	3	-6	3	-3
7	3	3	3	3	-3	7	7	3	2
-3	-7	-7	1	-2	2	-1	-3	3	7

21	22	23	24	25	26	27	28	29	30
9	3	5	2	13	8	11	1	15	7
-1	1	3	1	7	7	-7	10	-7	-1
3	7	7	7	-3	3	5	-7	3	2
2	3	3	-3	-7	7	3	-1	-7	3

Exercises: Big Friends 3 & 7

1	2	3	4	5	6	7	8	9	10
1	9	6	4	7	8	5	2	3	11
2	-1	2	7	2	7	13	1	7	-7
7	7	7	-3	3	-5	7	7	8	5
-3	-5	1	1	-1	-3	-5	-3	7	7
2	-3	-7	7	-7	1	-3	1	1	-1

11	12	13	14	15	16	17	18	19	20
6	4	7	5	3	9	1	8	2	10
3	5	1	3	1	7	7	1	2	-3
7	3	7	7	7	-5	7	7	7	2
-5	-1	3	-5	-3	-3	3	-5	-3	7
-3	-7	3	-3	1	7	7	-3	7	-1

21	22	23	24	25	26	27	28	29	30
9	3	8	4	2	6	5	7	1	10
3	7	-2	7	6	10	4	-6	10	-7
-2	5	3	1	7	-7	7	3	-3	6
-7	-7	7	-3	-5	3	-5	7	7	7
1	-2	-6	-5	-3	-1	-3	-3	-5	-1

Exercises: Big Friends 1&9, 2&8 3&7

1	2	3	4	5	6	7	8	9	10
3	7	1	2	5	9	4	8	6	7
8	8	9	2	1	3	7	3	9	-5
-7	-9	-7	7	9	-8	-8	-9	-8	9
9	2	8	-2	-8	-1	7	6	3	-7

11	12	13	14	15	16	17	18	19	20
8	5	9	6	7	3	4	2	1	8
-2	2	8	-5	-1	7	8	6	7	8
3	3	-6	9	9	-2	-9	7	8	-7
3	-1	-7	-3	-8	3	7	-9	-9	8

21	22	23	24	25	26	27	28	29	30
3	4	7	2	8	1	5	4	6	9
1	-2	-5	8	-1	3	4	-1	2	7
8	9	8	-1	3	7	3	7	2	-8
-9	-7	-3	3	-9	-2	-9	-2	-7	-6

Exercises: Big Friends 1&9, 2&8 3&7

1	2	3	4	5	6	7	8	9	10
5	2	8	3	1	6	7	9	4	3
4	1	9	7	5	2	2	-6	9	8
1	9	-6	-1	9	7	2	7	-3	-7
-3	-8	-7	7	-8	-8	-8	-2	-1	8
8	7	9	-8	-2	9	5	9	8	-9

11	12	13	14	15	16	17	18	19	20
9	6	4	7	5	8	3	4	1	2
7	9	6	-6	3	8	9	8	7	5
-8	-7	-2	9	7	9	-8	-9	7	9
7	8	7	-3	-9	-7	7	7	-8	-7
-8	-6	-8	9	2	8	-8	-1	9	8

21	22	23	24	25	26	27	28	29	30
3	5	2	4	6	7	1	9	4	8
5	1	6	-2	1	-1	8	-1	-3	7
8	9	7	9	8	9	7	7	9	-9
9	-7	-8	-7	-9	-8	2	-8	-7	2
-7	8	9	8	-1	2	8	9	9	7

Formulas for Big Friends 4 & 6

+ 4 = - 6 + 10

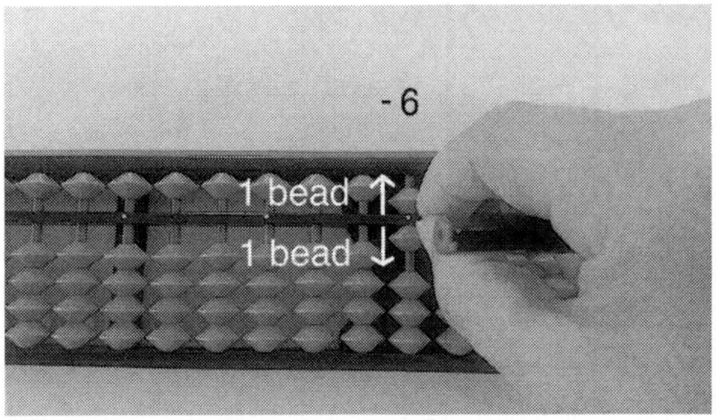

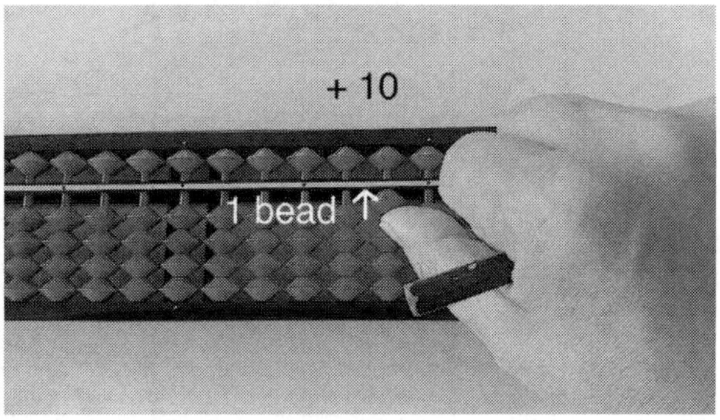

Formula **+ 4 = - 6 + 10** Exercises

1	2	3	4	5	6	7	8
6	7	8	9	16	17	18	19
4	4	4	4	4	4	4	4

1	2	3	4	5	6	7	8	9	10
2	3	3	5	9	6	8	7	6	5
5	6	5	4	-1	2	1	2	1	1
4	4	4	4	4	4	4	4	4	4

Formulas for Big Friends 4 & 6

- 4 = - 10 + 6

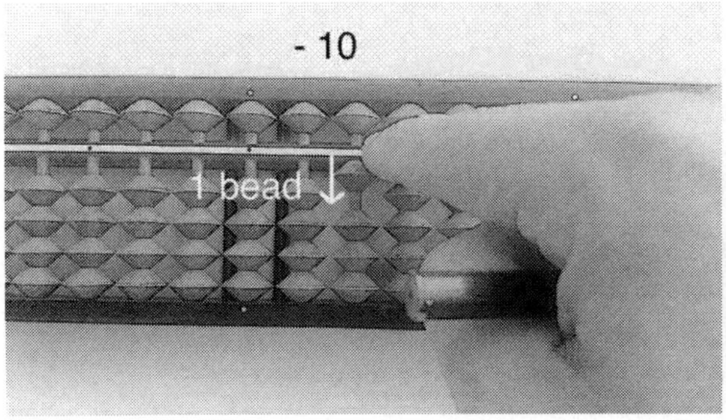

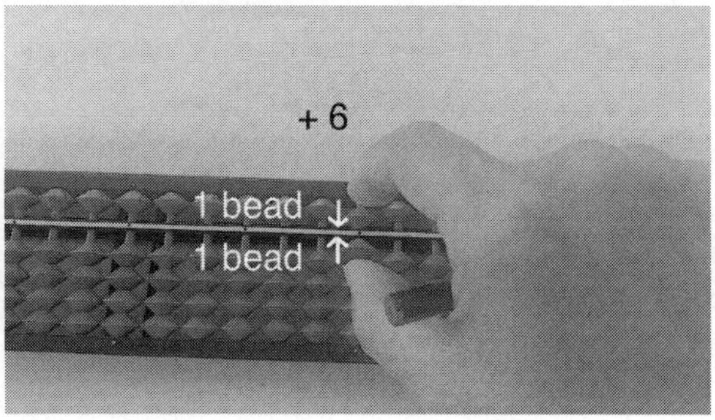

Formula - 4 = - 10 + 6 Exercises

1	2	3	4	5	6	7	8
10 -4	11 -4	12 -4	13 -4	20 -4	21 -4	22 -4	23 -4

1	2	3	4	5	6	7	8	9	10
10 1 -4	11 -1 -4	12 -1 -4	13 -2 -4	14 -4 -4	16 -5 -4	17 -5 -4	18 -6 -4	14 -2 -4	19 -6 -4

Formulas for Big Friends 4 & 6

+ 6 = - 4 + 10

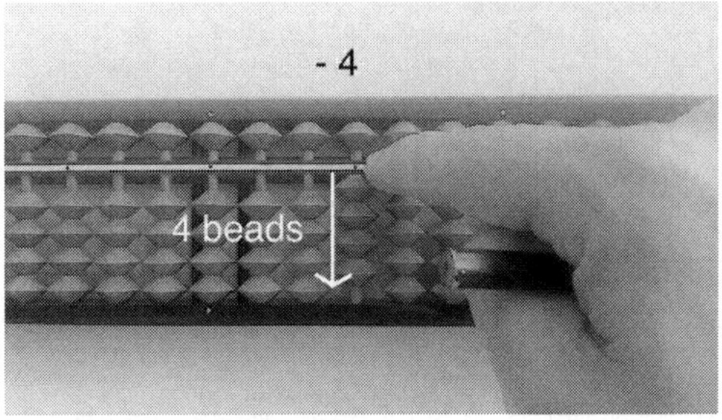

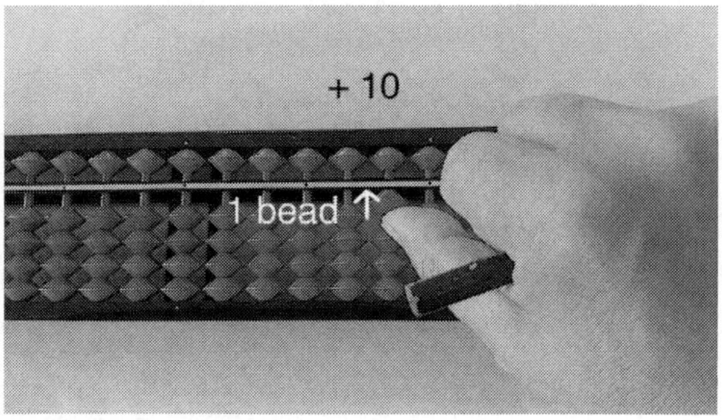

Formula **+ 6 = - 4 + 10** Exercises

1	2	3	4	5	6	7	8
4	9	14	19	24	29	34	39
6	6	6	6	6	6	6	6

1	2	3	4	5	6	7	8	9	10
4	5	3	6	2	3	1	7	8	1
5	4	6	3	2	1	3	2	1	8
6	6	6	6	6	6	6	6	6	6

Formulas for Big Friends 4 & 6

- 6 = - 10 + 4

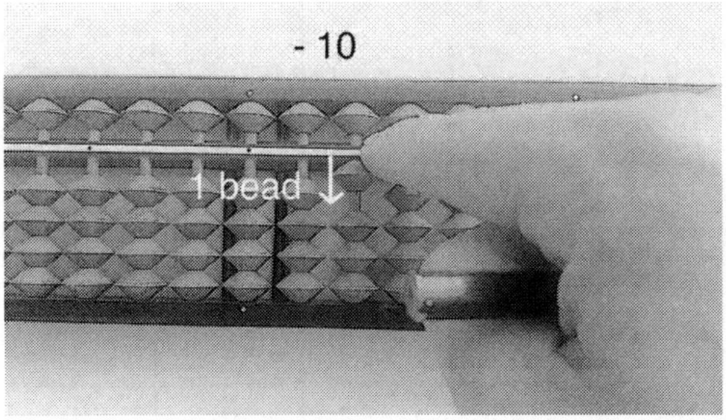

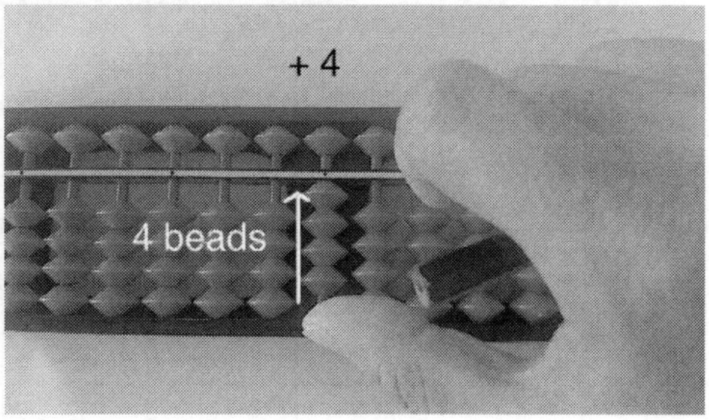

Formula - **6** = - **10** + **4** Exercises

1	2	3	4	5	6	7	8
10 -6	15 -6	20 -6	25 -6	30 -6	35 -6	40 -6	45 -6

1	2	3	4	5	6	7	8	9	10
10 5 -6	16 -1 -6	17 -2 -6	13 -3 -6	14 -4 -6	16 -6 -6	17 -7 -6	18 -3 -6	12 -2 -6	11 -1 -6

Exercises: Big Friends 4 & 6

1	2	3	4	5	6	7	8	9	10
4	2	9	1	5	7	3	8	6	10
6	2	4	3	1	4	6	4	2	-4
-4	6	-3	6	4	-1	6	-2	4	3
1	-4	-6	-4	-6	-6	-5	-6	2	6

11	12	13	14	15	16	17	18	19	20
5	8	14	7	2	6	1	4	3	9
4	1	6	-1	6	4	7	5	5	6
4	4	-4	4	4	7	4	6	4	-5
5	-2	3	-6	7	4	6	2	5	-4

21	22	23	24	25	26	27	28	29	30
6	1	13	5	3	8	2	9	4	7
3	15	-4	3	1	-2	7	-2	-2	2
4	4	6	4	6	4	6	4	6	4
1	-6	-5	-2	-4	-6	2	5	4	-3

Exercises: Big Friends 4 & 6

1	2	3	4	5	6	7	8	9	10
6	4	1	5	8	3	7	2	9	10
4	6	7	3	4	6	1	6	-2	-4
6	4	4	4	-2	6	4	4	4	3
4	6	-2	6	-6	-5	5	6	-1	6
6	4	-6	4	5	-4	4	4	-6	2

11	12	13	14	15	16	17	18	19	20
3	5	7	6	9	2	4	8	1	16
5	1	-1	2	4	5	5	1	3	4
4	4	4	4	-3	4	6	4	6	-6
-2	-6	8	-2	-6	-1	2	5	-4	5
-6	5	4	-6	5	-6	4	4	1	4

21	22	23	24	25	26	27	28	29	30
9	4	2	3	8	5	1	7	6	14
6	10	2	1	-2	2	5	2	3	6
2	6	6	6	4	4	4	4	6	-4
4	-4	-4	-4	-6	-1	-6	-3	-5	2
5	2	3	1	5	-6	-2	-6	-4	4

Exercises: Big Friends 1&9, 2&8, 3&7, 4&6

1	2	3	4	5	6	7	8	9	10
8	3	5	7	2	9	4	1	6	2
2	7	2	9	9	6	7	9	2	8
-1	-6	9	-8	-7	-5	-3	-4	7	-4
6	8	-7	7	6	-2	2	9	-9	9

11	12	13	14	15	16	17	18	19	20
4	2	6	8	5	1	3	7	9	2
6	5	9	9	4	7	8	2	7	2
-1	3	-8	9	6	7	-1	6	9	6
6	-6	9	-7	-9	-6	-4	-8	-8	-4

21	22	23	24	25	26	27	28	29	30
7	5	3	4	9	6	2	1	8	5
8	1	9	8	6	3	7	8	4	3
-6	4	8	-9	-7	6	4	4	-8	2
7	-6	-4	7	8	-9	-9	7	6	-4

Exercises: Big Friends 1&9, 2&8, 3&7, 4&6

1	2	3	4	5	6	7	8	9	10
7	3	6	1	8	2	4	9	5	3
3	8	4	5	4	8	6	4	2	9
-4	-7	-7	4	-9	-6	-7	7	9	-8
9	6	8	-6	7	7	8	-4	-8	6
-6	-4	-9	9	-4	9	-9	9	7	-4

11	12	13	14	15	16	17	18	19	20
4	5	7	2	1	3	6	8	9	4
7	4	9	2	6	8	3	7	3	5
9	6	-8	6	8	5	4	-9	8	3
-6	-8	7	-8	-5	4	7	4	-4	-8
8	9	-8	9	-4	-6	-9	-6	-7	7

21	22	23	24	25	26	27	28	29	30
1	6	3	8	5	9	2	7	4	2
9	3	6	9	3	9	1	3	6	9
-6	8	2	8	7	-7	7	-6	-8	9
7	-7	-7	-6	-6	9	-4	7	9	-6
-8	6	6	3	4	-6	9	-9	-7	7

Formulas for Big Friends 5 & 5

+ 5 = - 5 + 10

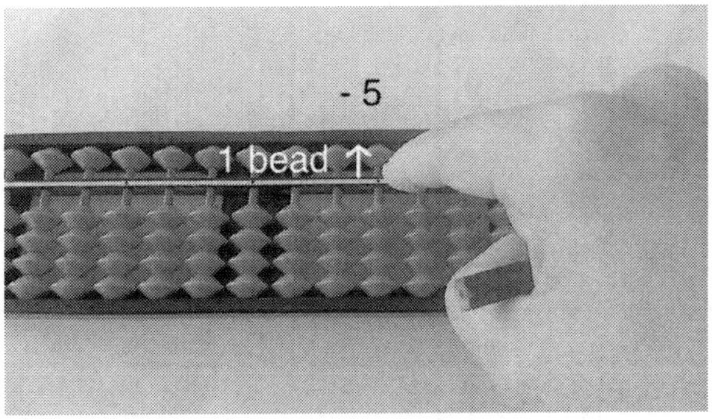

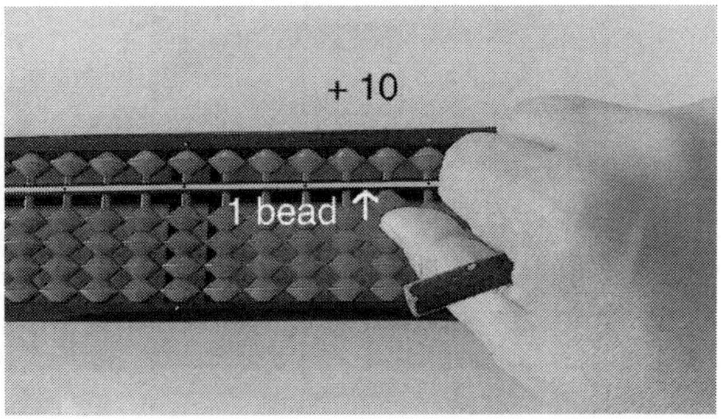

- 5 = - 10 + 5

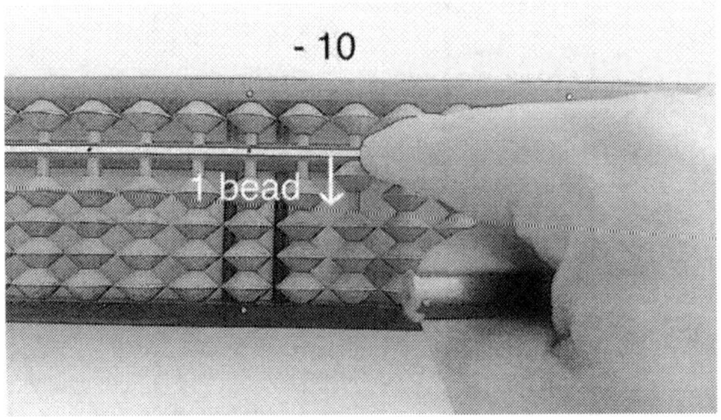

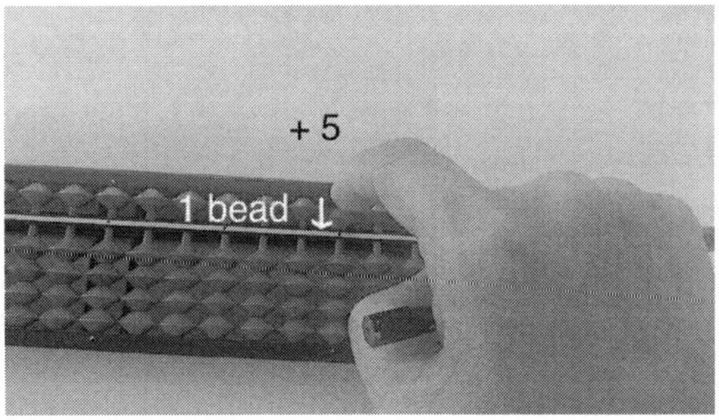

Exercises: Big Friends 1&9, 2&8, 3&7, 4&6, 5

1	2	3	4	5	6	7	8	9	10
5	8	3	6	4	2	9	7	1	4
5	2	7	9	5	8	-5	9	2	8
-6	-5	-5	-7	-4	5	6	-1	7	-5
7	1	2	2	5	5	-5	5	-5	8
-8	4	3	-5	-1	-6	2	-4	3	-7

11	12	13	14	15	16	17	18	19	20
7	4	1	8	3	6	2	5	9	8
5	5	5	-3	5	5	1	3	-7	5
8	5	9	5	5	9	7	2	8	-9
-9	6	-6	-4	7	-8	-5	-5	-5	7
-5	-5	5	5	-5	-8	2	4	3	-4

21	22	23	24	25	26	27	28	29	30
5	8	2	4	7	1	5	3	1	9
1	-2	9	6	8	7	2	9	6	-6
4	4	-5	-3	5	5	5	6	3	7
-7	-5	4	5	-6	7	-8	5	-5	-5
8	3	-5	-8	8	-4	6	-4	1	4

Chapter 6: Formulas for Little Friends

Formulas for Little Friends 1 & 4

+ 1 = + 5 – 4

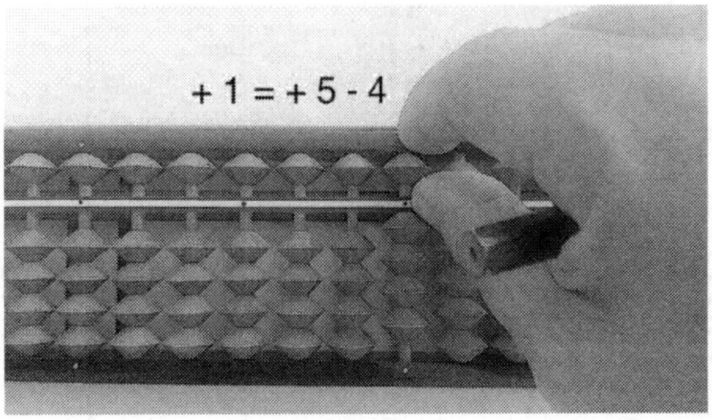

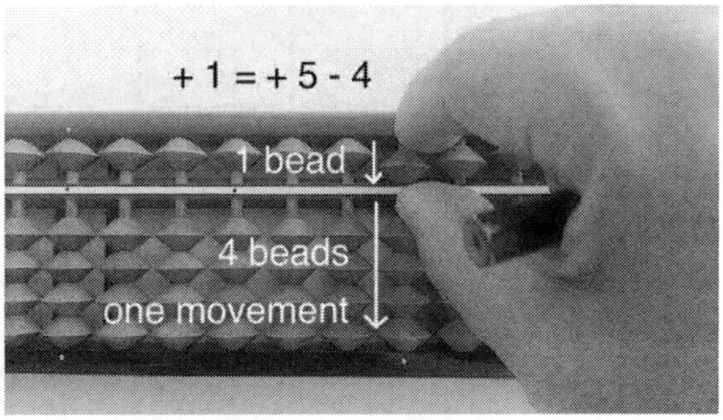

Formula + 1 = + 5 - 4 Exercises

1	2	3	4	5	6	7	8
4	14	24	34	44	54	64	74
1	1	1	1	1	1	1	1

1	2	3	4	5	6	7	8	9	10
2	1	3	10	11	2	1	12	21	23
2	3	1	4	3	12	13	2	3	1
1	1	1	1	1	1	1	1	1	1

Formulas for Little Friends 1 & 4

- 1 = - 5 + 4

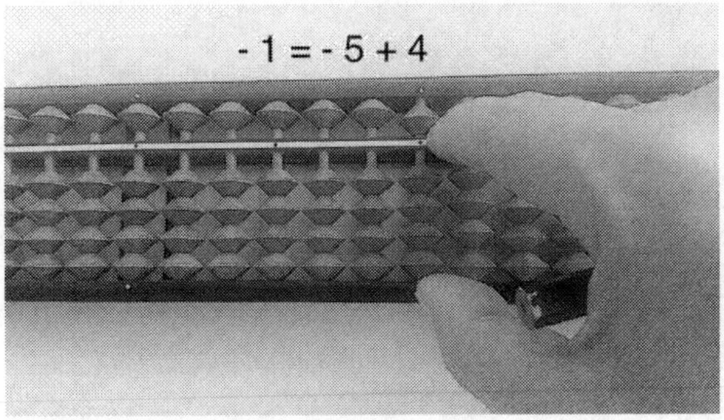

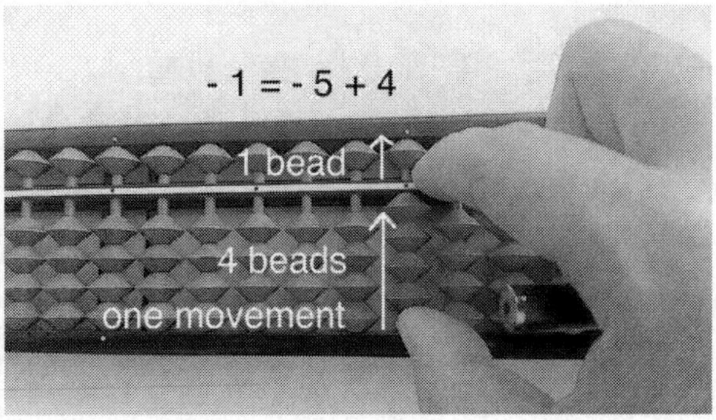

Formula **- 1 = - 5 + 4** Exercises

1	2	3	4	5	6	7	8
5 -1	15 -1	25 -1	35 -1	45 -1	55 -1	65 -1	75 -1

1	2	3	4	5	6	7	8	9	10
8 -3 -1	6 -1 -1	7 -2 -1	9 -4 -1	16 -1 -1	17 2 -1	18 -3 -1	19 -4 -1	26 -1 -1	28 -3 -1

Formulas for Little Friends 1 & 4

+ 4 = + 5 – 1

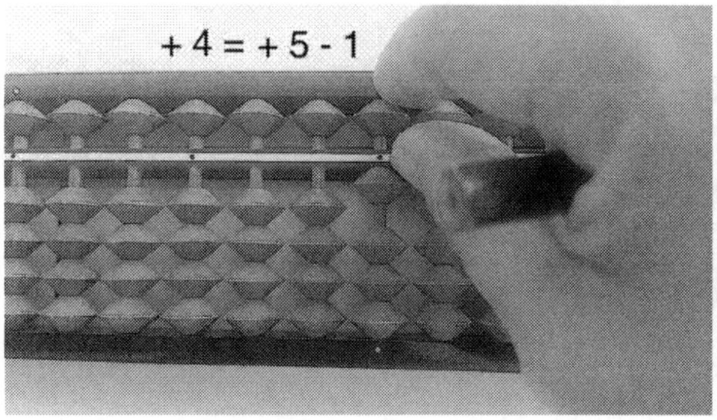

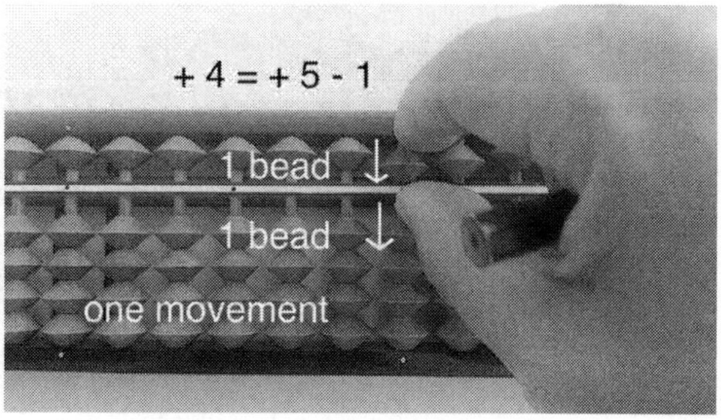

Formula **+ 4 = + 5 - 1** Exercises

1	2	3	4	5	6	7	8
1	2	3	4	11	12	13	14
4	4	4	4	4	4	4	4

1	2	3	4	5	6	7	8	9	10
2	1	1	6	7	8	9	11	12	13
2	3	1	-5	-5	-5	-5	2	-1	-1
4	4	4	4	4	4	4	4	4	4

Formulas for Little Friends 1 & 4

- 4 = - 5 + 1

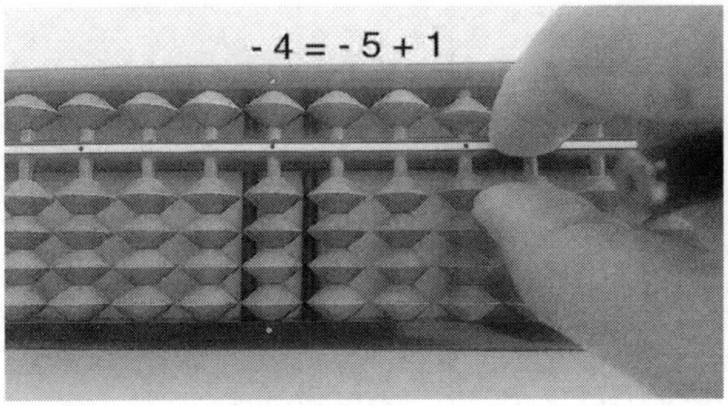

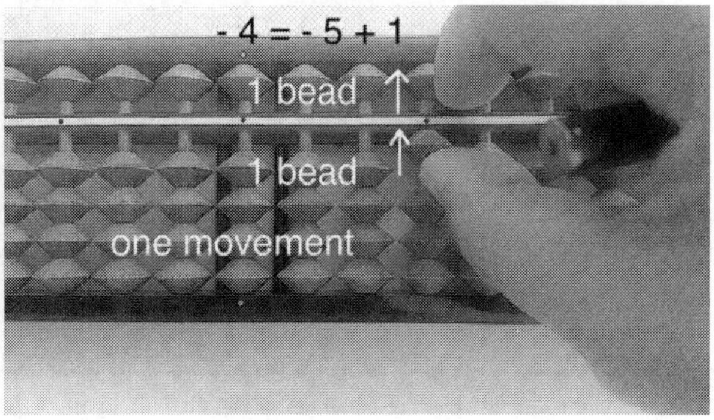

Formula - 4 = - 5 + 1 Exercises

1	2	3	4	5	6	7	8
5 -4	6 -4	7 -4	8 -4	15 -4	16 -4	17 -4	18 -4

1	2	3	4	5	6	7	8	9	10
9 -4 -4	6 -1 -4	7 -1 -4	9 -2 -4	8 -3 -4	7 -2 -4	8 -2 -4	9 -1 -4	9 -3 -4	8 -1 -4

Exercises: Little Friends 1 & 4

1	2	3	4	5	6	7	8	9	10
1	4	6	9	3	7	2	5	8	6
4	4	-5	-4	4	-4	2	-4	-4	-4
-1	-5	4	-1	-2	1	1	3	1	5
5	4	-1	4	-1	1	-4	1	-4	-4

11	12	13	14	15	16	17	18	19	20
7	3	8	2	4	1	9	7	5	8
1	5	-3	4	1	6	-5	-2	-1	-2
-4	-4	-1	-5	2	-4	1	-4	1	-4
1	1	1	4	-4	1	-4	6	-1	1

21	22	23	24	25	26	27	28	29	30
4	6	2	5	6	11	3	8	1	7
1	1	1	1	-1	4	-2	-5	3	-5
1	-4	4	-4	-4	-1	4	4	1	4
-4	1	1	2	2	5	-1	2	4	1

Exercises: Little Friends 1 & 4 and Big Friends

1	2	3	4	5	6	7	8	9	10
5	3	7	1	4	8	2	9	6	1
-1	1	4	4	6	-7	4	1	5	6
6	1	4	5	-5	4	4	-5	4	8
-4	5	-6	-6	-1	5	-6	-4	-7	-4
-1	-6	5	1	6	-1	1	9	8	9

11	12	13	14	15	16	17	18	19	20
2	8	6	3	7	1	5	9	4	5
5	7	9	-2	8	8	5	2	7	-4
4	-1	-4	4	-1	5	-6	4	4	6
4	8	9	5	6	1	1	-1	-1	-4
-6	-9	-4	-2	-4	-4	5	6	4	7

21	22	23	24	25	26	27	28	29	30
9	1	5	2	8	6	4	7	3	5
-5	7	4	2	-4	-4	8	-4	7	-1
1	7	6	1	1	8	4	8	-2	7
4	-4	-1	5	5	-5	-1	-5	-4	4
6	9	-5	-9	-4	-1	-1	-4	1	-1

Formulas for Little Friends 2 & 3

+ 2 = + 5 – 3

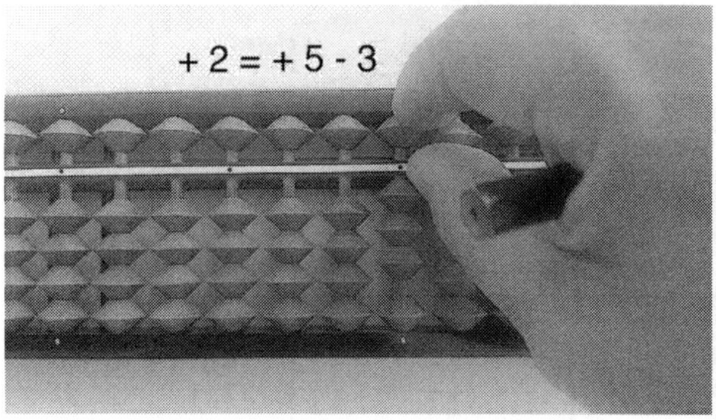

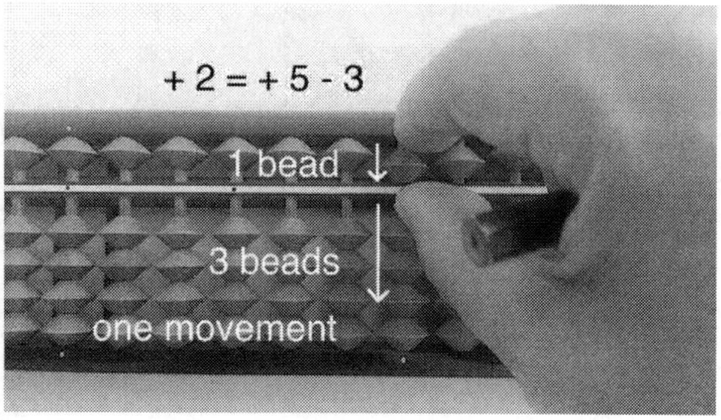

Formula **+ 2 = + 5 - 3** Exercises

1	2	3	4	5	6	7	8
3 2	4 2	13 2	14 2	23 2	24 2	33 2	34 2

1	2	3	4	5	6	7	8	9	10
1 2 2	8 -5 2	9 -6 2	4 -1 2	2 1 2	9 -5 2	2 2 2	3 1 2	1 3 2	11 2 2

Formulas for Little Friends 2 & 3

- 2 = - 5 + 3

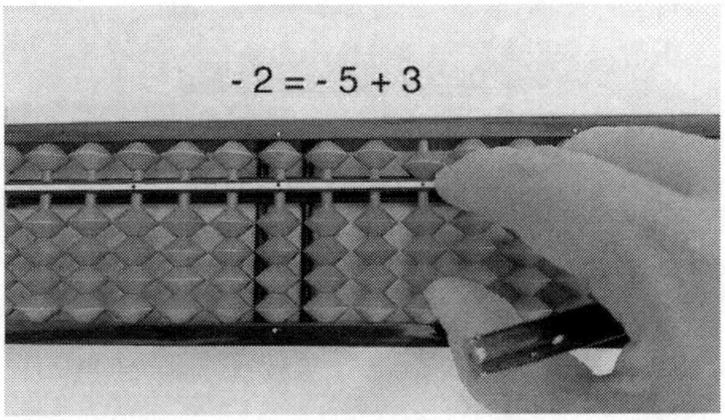

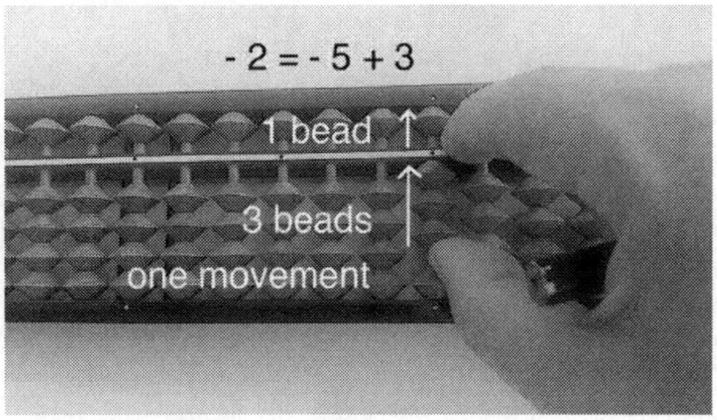

Formula **- 2 = - 5 + 3** Exercises

1	2	3	4	5	6	7	8
5 -2	6 -2	15 -2	16 -2	25 -2	26 -2	35 -2	36 -2

1	2	3	4	5	6	7	8	9	10
8 -3 -2	6 -1 -2	7 -2 -2	9 -4 -2	8 -2 -2	6 -2 -2	7 -1 -2	9 -3 -2	5 1 -2	1 5 -2

Formulas for Little Friends 2 & 3

+ 3 = + 5 – 2

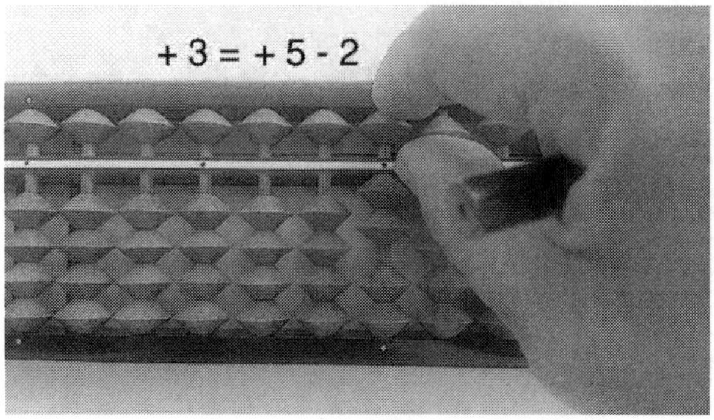

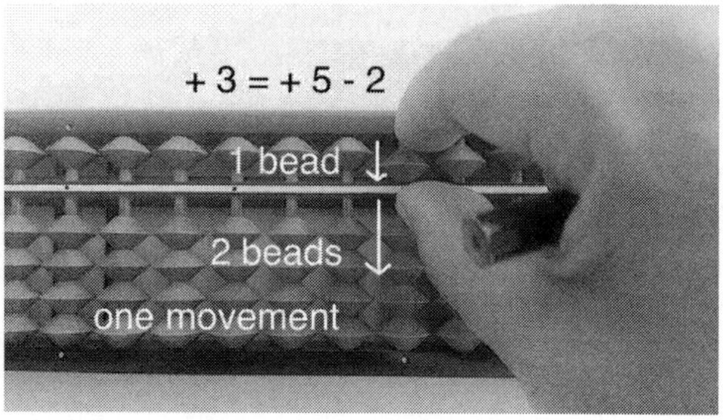

Formula **+ 3 = + 5 - 2** Exercises

1	2	3	4	5	6	7	8
2	3	4	12	13	14	22	23
3	3	3	3	3	3	3	3

1	2	3	4	5	6	7	8	9	10
1	7	9	4	8	3	2	8	9	2
1	-5	-7	-1	-6	1	1	-5	-6	2
3	3	3	3	3	3	3	3	3	3

Formulas for Little Friends 2 & 3

- 3 = - 5 + 2

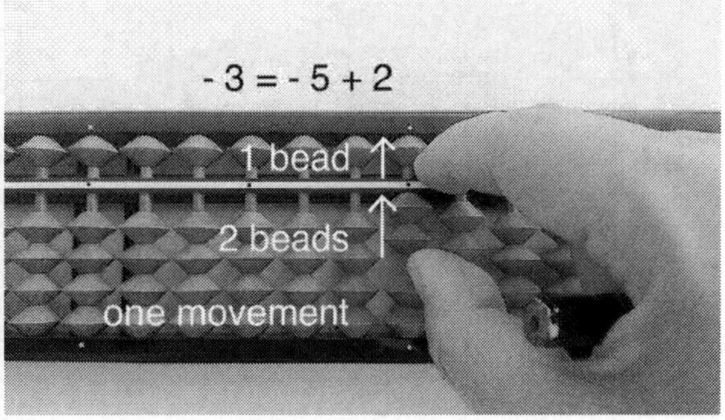

Formula **- 3 = - 5 + 2** Exercises

1	2	3	4	5	6	7	8
5 -3	6 -3	7 -3	15 -3	16 -3	17 -3	25 -3	26 -3

1	2	3	4	5	6	7	8	9	10
8 -3 -3	6 -1 -3	7 -1 -3	9 -4 -3	8 -2 -3	7 -2 -3	9 -2 -3	9 -3 -3	8 -1 -3	6 -3 -3

Exercises: Little Friends 2 & 3

1	2	3	4	5	6	7	8	9	10
2	5	4	7	1	3	6	8	9	1
3	-3	2	-2	3	2	-1	-5	-7	1
-2	1	-3	-3	2	1	-2	2	3	3
1	2	2	2	-3	-3	1	-3	-2	-2

11	12	13	14	15	16	17	18	19	20
8	3	6	2	5	4	1	8	7	15
-6	2	1	1	-2	3	2	-3	-5	-3
3	2	-3	2	3	-1	2	-2	3	2
-2	-3	5	-3	1	-2	-3	3	-2	2

21	22	23	24	25	26	27	28	29	30
9	4	5	7	3	6	7	1	2	18
-4	-2	2	-2	1	-1	-2	3	5	-13
-3	3	-3	-3	3	-3	-2	2	-3	-3
1	-2	2	1	1	2	3	-3	2	6

Exercises: Little Friends 2 & 3 and Big Friends

1	2	3	4	5	6	7	8	9	10
3	5	9	1	4	8	2	6	3	7
2	-2	3	2	3	-6	3	-2	3	-3
5	7	3	2	-5	3	3	6	5	6
-6	-6	-2	5	3	5	9	-7	-8	-8
7	8	9	-4	5	-1	-8	2	2	3

11	12	13	14	15	16	17	18	19	20
8	5	3	7	1	4	6	2	9	3
5	-2	3	8	6	2	-2	1	3	9
2	3	-2	-3	5	-3	6	3	3	3
-3	4	5	-9	3	7	-5	5	-9	-2
-9	-8	6	2	-2	-9	-3	-9	-2	-9

21	22	23	24	25	26	27	28	29	30
4	9	6	8	5	2	7	1	3	4
3	4	-3	4	-3	2	9	5	2	8
5	2	2	-9	9	6	-2	-2	4	3
3	-3	5	2	-8	-5	-5	8	8	-2
-6	-8	-7	-3	2	-2	4	3	-3	7

Exercises: Big Friends and Little Friends

1	2	3	4	5	6	7	8	9	10
6	2	8	5	3	9	4	1	7	6
5	3	7	-2	4	6	1	9	-3	-2
-7	5	-3	8	3	-4	-3	-5	7	1
1	-8	8	4	-6	-7	8	-1	-9	5
-3	3	-9	-7	1	1	-6	6	8	-4
11	12	13	14	15	16	17	18	19	20
5	4	6	9	2	7	1	8	4	3
5	9	9	7	9	5	4	8	8	7
-8	2	-1	4	4	-8	5	-3	3	-5
3	-7	6	-8	-3	1	-6	7	-1	-4
-2	2	-8	3	8	-2	1	-9	6	9
21	22	23	24	25	26	27	28	29	30
8	3	7	2	5	4	6	9	1	3
-4	2	9	8	-4	7	-3	9	5	2
6	-1	-4	-7	9	-8	2	-4	-2	-1
-5	6	-8	2	-8	2	5	1	7	7
-1	-9	6	-1	3	-1	-9	-2	9	-8

Exercises: Big Friends and Little Friends

1	2	3	4	5	6	7	8	9	10
5	2	8	4	1	3	7	9	2	6
2	1	-4	2	6	2	5	3	2	4
3	7	9	5	8	-1	3	-8	7	-9
-4	-6	2	-8	-9	7	-4	1	-8	4
-2	1	-7	7	-4	-8	-9	-2	2	-3

11	12	13	14	15	16	17	18	19	20
9	3	7	1	5	8	2	4	6	3
2	4	3	8	5	8	3	3	5	1
4	8	-6	6	-8	-3	-4	3	-7	7
-8	-9	8	-3	3	7	9	-5	2	4
-3	-2	3	2	-1	-9	-6	-1	4	-6

21	22	23	24	25	26	27	28	29	30
6	2	4	3	8	5	1	7	9	5
-2	8	1	7	-5	-1	7	8	8	-4
1	-5	5	-2	7	3	7	-6	-3	9
-3	-3	-6	-4	-5	3	-4	3	6	-6
8	8	1	1	-1	-8	-9	3	-4	2

Chapter 7: Formulas for Mixed Friends

Formulas for Mixed Friend + 6

+ 6 = - 5 + 1 + 10

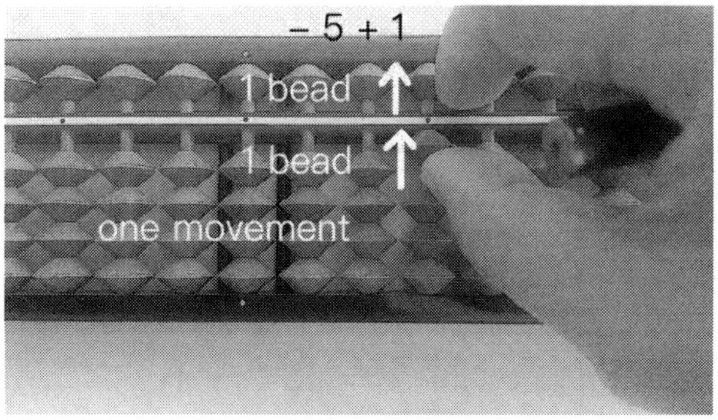

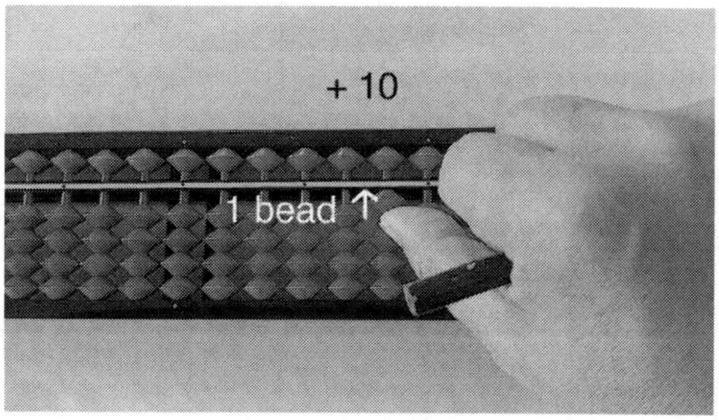

Formula **+ 6 = - 5 + 1 + 10** Exercises

1	2	3	4	5	6	7	8
5	6	7	8	15	16	17	18
6	6	6	6	6	6	6	6

1	2	3	4	5	6	7	8	9	10
5	7	7	9	8	5	2	9	7	9
1	-2	-1	-1	-2	2	6	-3	1	-2
6	6	6	6	6	6	6	6	6	6

Formulas for Mixed Friend - 6

- 6 = - 10 + 5 - 1

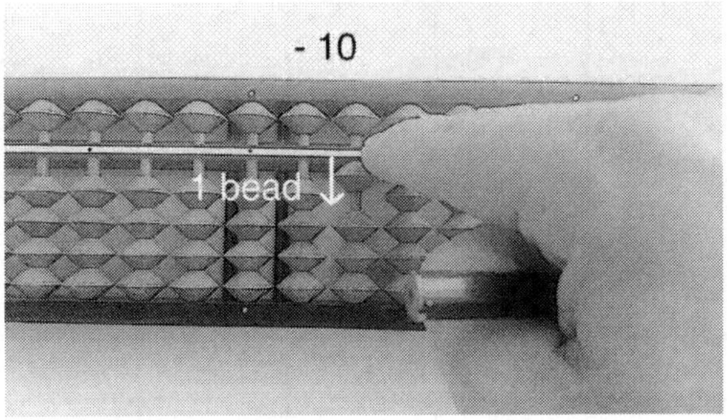

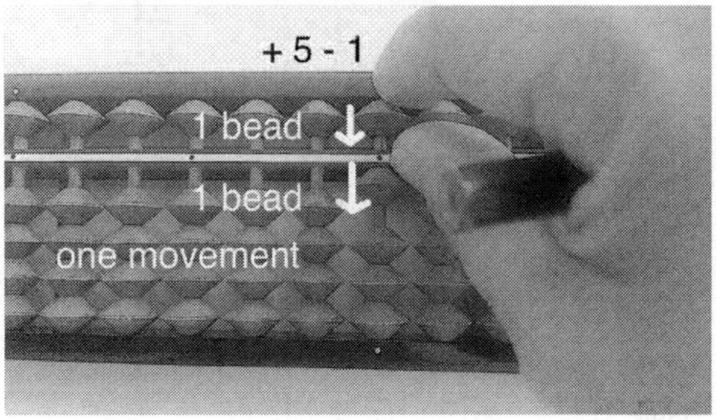

Formula **- 6 = - 10 + 5 - 1** Exercises

1	2	3	4	5	6	7	8
11 -6	12 -6	13 -6	14 -6	21 -6	22 -6	23 -6	24 -6

1	2	3	4	5	6	7	8	9	10
10 1 -6	11 1 -6	12 2 -6	13 -1 -6	12 1 -6	11 2 -6	14 -1 -6	14 -3 -6	13 -2 -6	11 3 -6

Exercises: Mixed Friend 6

1	2	3	4	5	6	7	8	9	10
5	7	11	2	8	14	9	7	3	12
6	6	-6	6	6	-6	-2	-2	5	-6
6	1	11	6	-2	-2	6	6	6	2
6	-4	-6	-4	6	3	-2	3	-1	-3

11	12	13	14	15	16	17	18	19	20
6	8	3	5	1	2	14	4	7	9
6	-1	15	2	10	5	-6	-2	6	-1
2	6	6	6	-6	6	-2	5	5	6
-1	1	5	1	2	5	1	6	6	5

21	22	23	24	25	26	27	28	29	30
6	7	14	13	11	1	12	3	9	11
1	-1	-1	-6	2	6	1	10	-3	-6
6	6	-6	1	1	6	-6	-6	6	2
5	1	1	-2	-6	1	2	1	1	1

Exercises: Mixed Friend 6 and Big Friends + Little Friends

1	2	3	4	5	6	7	8	9	10
2	3	8	9	1	4	7	6	5	8
3	9	7	8	9	7	6	6	6	8
6	-6	6	-5	2	-6	8	9	-4	6
-4	2	-3	-6	-6	2	-9	-7	6	-9
5	2	6	4	2	4	-3	-4	2	3

11	12	13	14	15	16	17	18	19	20
5	7	1	5	3	2	4	8	9	6
2	4	7	-2	4	6	3	6	4	5
6	-6	6	9	6	6	6	1	-6	9
3	3	-2	-6	2	-3	-2	6	3	-4
6	2	4	2	6	-5	-8	3	-8	6

21	22	23	24	25	26	27	28	29	30
2	8	3	14	7	5	1	4	7	9
4	8	3	-3	3	3	12	3	1	3
6	6	6	-6	5	6	-6	6	3	-6
9	3	4	-1	6	1	-4	2	-6	4
-7	6	-2	2	-1	6	3	6	2	5

Formulas for Mixed Friend + 7

+ 7 = - 5 + 2 + 10

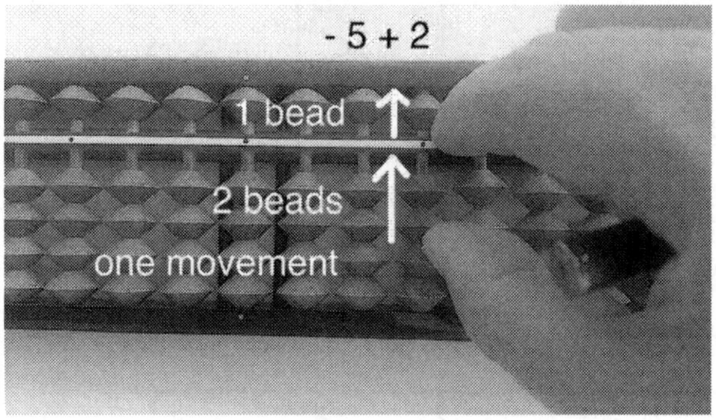

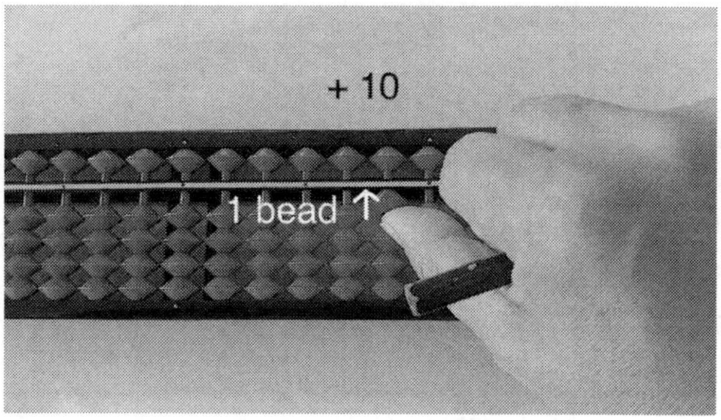

Formula **+ 7 = - 5 + 2 + 10** Exercises

1	2	3	4	5	6	7	8
5 7	6 7	7 7	15 7	16 7	17 7	25 7	26 7

1	2	3	4	5	6	7	8	9	10
5 1 7	7 -2 7	8 -3 7	9 -4 7	8 -2 7	5 2 7	1 5 7	9 -3 7	7 -1 7	8 -1 7

Formulas for Mixed Friends - 7

- 7 = - 10 + 5 - 2

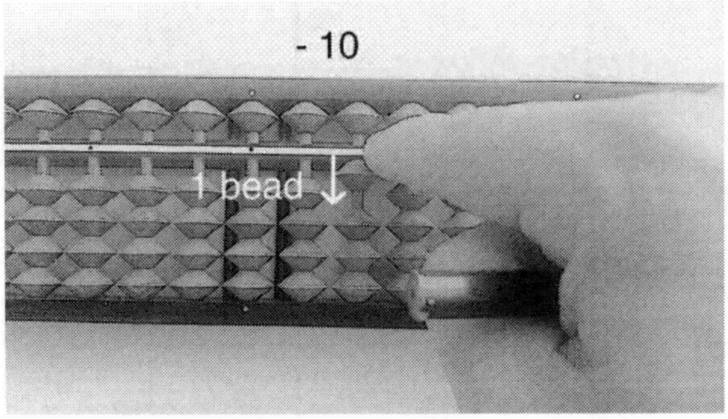

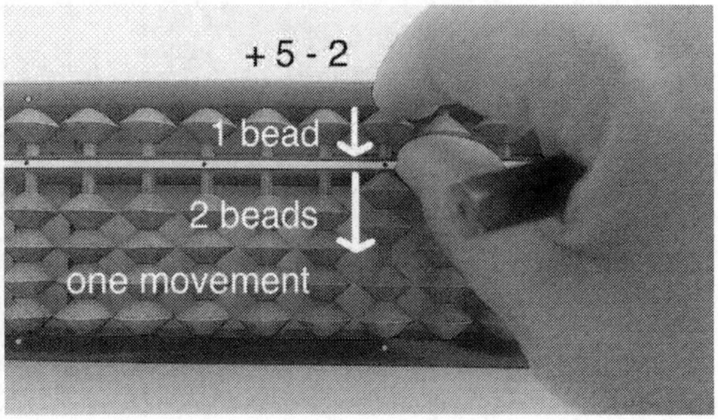

Formula - **7** = - **10** + **5** - **2** Exercises

1	2	3	4	5	6	7	8
12 -7	13 -7	14 -7	22 -7	23 -7	24 -7	32 -7	33 -7

1	2	3	4	5	6	7	8	9	10
10 2 -7	11 1 -7	12 2 -7	13 -1 -7	12 1 -7	11 2 -7	14 -1 -7	14 -2 -7	10 3 -7	11 3 -7

Exercises: Mixed Friend 7

1	2	3	4	5	6	7	8	9	10
5	12	6	7	2	3	1	4	8	13
7	-7	7	7	5	10	5	-2	-2	-7
5	1	1	-2	7	-7	7	5	7	1
7	3	-4	-2	-1	2	-2	7	-1	-5

11	12	13	14	15	16	17	18	19	20
9	12	2	1	8	7	6	3	4	14
-3	2	10	13	-1	-1	-1	-1	-3	-7
7	-7	-7	-7	7	7	7	5	5	1
1	-2	1	1	-2	1	-2	7	7	-5

21	22	23	24	25	26	27	28	29	30
1	4	7	9	14	8	6	2	3	15
6	10	-2	-4	-2	-3	7	11	11	3
7	-7	7	7	-7	7	-1	-7	-7	-5
-2	1	1	1	1	1	-2	2	1	-7

Exercises: Mixed Friend 7 and Big Friends + Little Friends

1	2	3	4	5	6	7	8	9	10
8	3	5	6	1	7	2	14	9	3
4	4	7	1	4	7	4	5	6	2
-7	7	3	7	7	3	7	-7	-3	7
2	-2	-6	1	-8	-4	-5	-7	-7	5
3	3	8	5	7	2	3	-4	4	-4

11	12	13	14	15	16	17	18	19	20
9	12	2	4	8	5	3	6	2	1
8	1	5	-2	-2	1	8	7	2	3
-5	-7	7	3	7	7	2	4	1	10
-7	-2	3	7	4	2	-7	-5	7	-7
-3	1	7	4	-4	-6	4	-8	-9	2

21	22	23	24	25	26	27	28	29	30
6	4	3	7	1	8	2	5	9	3
-4	3	10	-2	6	4	3	-2	4	3
3	7	-7	7	7	-7	7	9	-7	7
7	-1	5	-8	-2	3	-4	-7	4	-9
4	2	-7	7	4	-4	-4	4	-6	7

Formulas for Mixed Friend + 8

+ 8 = - 5 + 3 + 10

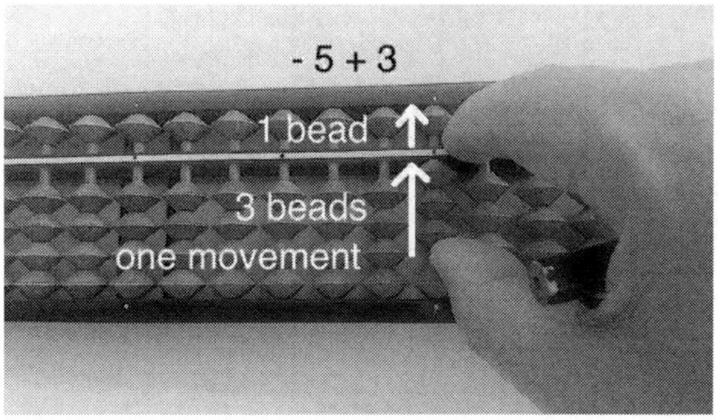

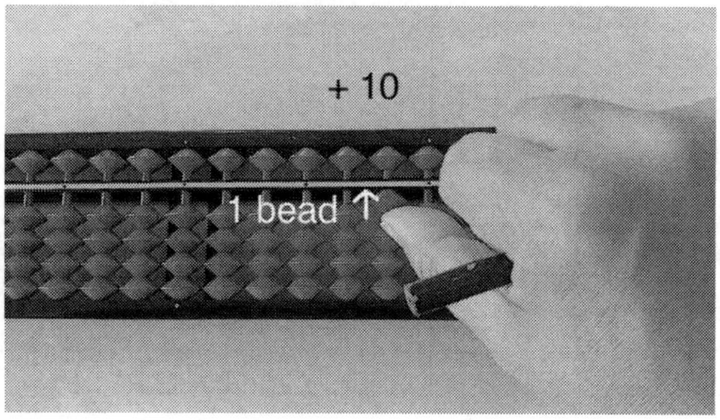

Formula **+ 8 = - 5 + 3 + 10** Exercises

1	2	3	4	5	6	7	8
5	6	15	16	25	26	35	36
8	8	8	8	8	8	8	8

1	2	3	4	5	6	7	8	9	10
5	7	8	9	17	15	18	9	7	8
1	-2	-3	-3	-2	1	-3	-4	-1	-2
8	8	8	8	8	8	8	8	8	8

Formulas for Mixed Friend - 8

- 8 = - 10 + 5 - 3

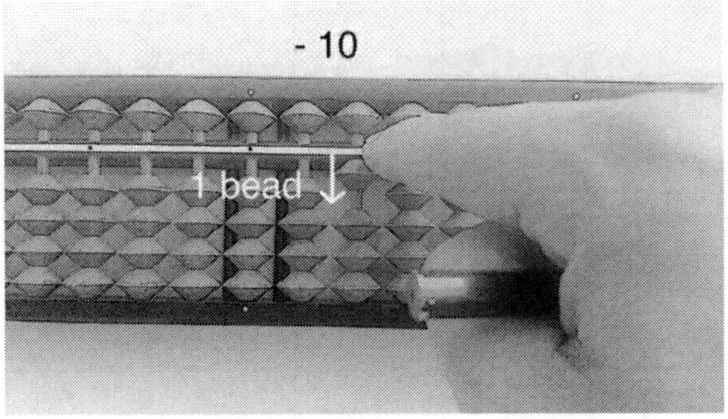

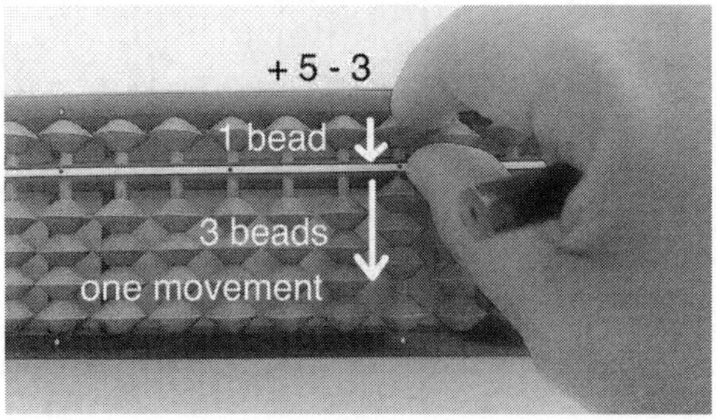

Formula **- 8 = - 10 + 5 - 3** Exercises

1	2	3	4	5	6	7	8
13	14	23	24	33	34	43	44
-8	-8	-8	-8	-8	-8	-8	-8

1	2	3	4	5	6	7	8	9	10
10	11	12	13	12	11	14	10	20	20
3	3	2	1	1	2	-1	4	3	4
-8	-8	-8	-8	-8	-8	-8	-8	-8	-8

Exercises: Mixed Friend 8

1	2	3	4	5	6	7	8	9	10
5	1	3	8	4	2	7	13	6	9
8	5	10	-2	10	5	-2	-8	8	-3
1	8	-8	8	-8	-1	8	1	-1	8
-4	-2	1	5	1	8	-1	-5	-3	-2

11	12	13	14	15	16	17	18	19	20
7	4	5	6	14	2	9	1	3	8
-1	-3	1	-1	-8	2	-4	2	-2	-3
8	5	8	8	1	10	8	10	13	8
-4	8	-2	-2	-2	-8	1	-8	-8	5

21	22	23	24	25	26	27	28	29	30
5	2	14	1	3	7	8	6	10	15
2	5	-1	12	1	-6	10	10	3	8
-1	-2	-8	-8	10	5	-5	8	-8	-10
8	8	1	2	-8	8	-8	-4	2	-3

Exercises: Mixed Friend 8 and Big Friends + Little Friends

1	2	3	4	5	6	7	8	9	10
6	4	2	1	9	3	6	9	5	15
8	2	3	1	-3	3	4	3	8	-1
1	8	8	11	8	8	-5	2	4	-8
8	-1	-4	-8	1	-5	8	-8	4	4
-4	2	7	2	-4	4	-5	-4	-3	-8

11	12	13	14	15	16	17	18	19	20
3	9	5	6	1	4	2	5	7	5
2	4	-1	4	5	1	7	1	-4	4
4	-8	9	3	8	8	4	8	10	4
4	-1	-8	-8	1	-4	-8	6	-8	-8
-8	3	-1	5	-7	6	-1	-7	2	5

21	22	23	24	25	26	27	28	29	30
1	8	17	2	5	3	9	4	6	3
3	-3	-4	6	8	2	3	9	3	1
9	8	-8	5	6	8	2	-8	9	9
-8	2	5	-8	7	2	-8	-1	-5	-8
-1	8	-8	4	-4	8	-6	6	-8	3

Formulas for Mixed Friend + 9

+ 9 = - 5 + 4 + 10

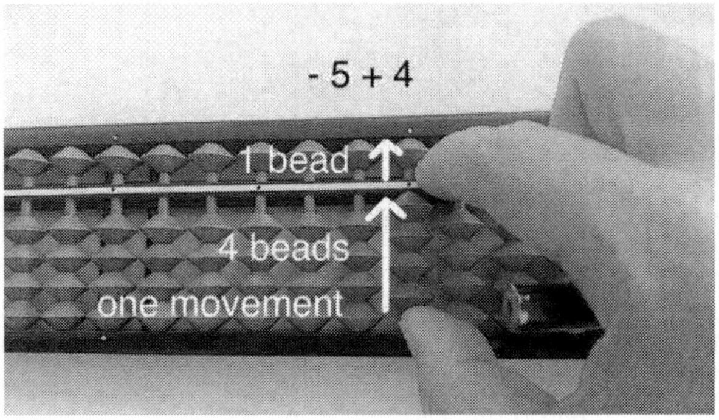

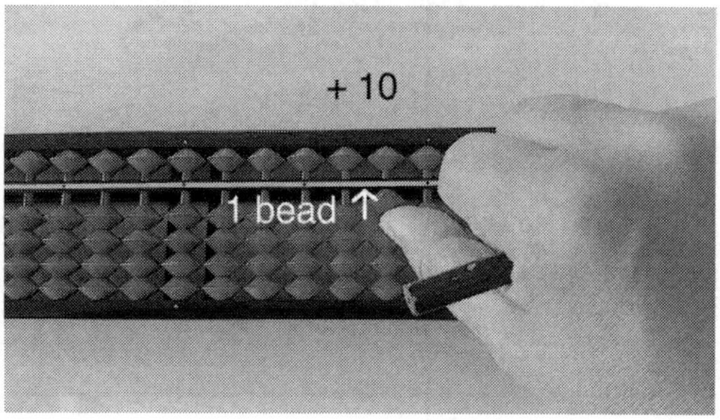

Formula **+ 9 = - 5 + 4 + 10** Exercises

1	2	3	4	5	6	7	8
5	15	25	35	45	55	65	75
9	9	9	9	9	9	9	9

1	2	3	4	5	6	7	8	9	10
10	5	20	5	30	5	40	5	50	5
5	10	5	20	5	30	5	40	5	50
9	9	9	9	9	9	9	9	9	9

Formulas for Mixed Friend - 9

- 9 = - 10 + 5 - 4

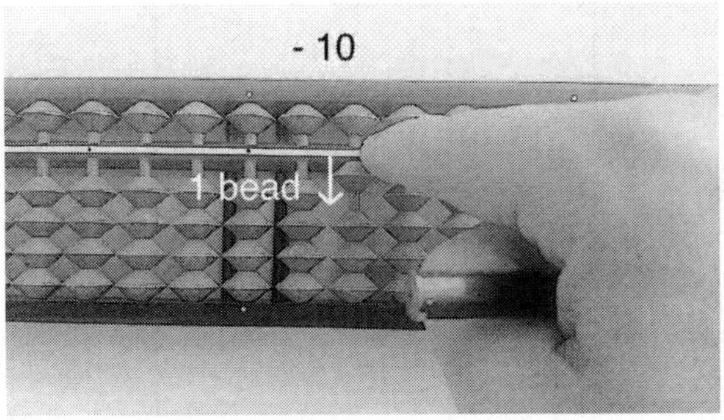

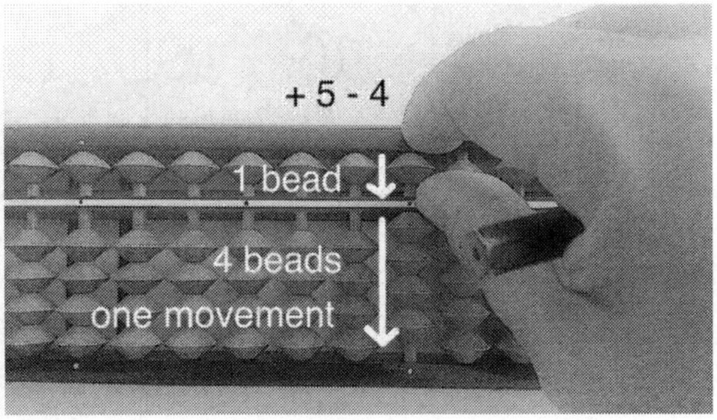

Formula **- 9 = - 10 + 5 - 4** Exercises

1	2	3	4	5	6	7	8
14 -9	24 -9	34 -9	44 -9	54 -9	64 -9	74 -9	84 -9

1	2	3	4	5	6	7	8	9	10
10 4 -9	11 3 -9	12 2 -9	13 1 -9	20 4 -9	21 3 -9	22 2 -9	23 1 -9	30 4 -9	31 3 -9

Exercises: Mixed Friend 9

1	2	3	4	5	6	7	8	9	10
5	14	1	10	2	8	4	3	6	7
9	-9	3	5	1	-3	15	11	-1	12
-2	2	10	9	11	9	-5	-9	9	-5
1	-1	-9	-4	-9	-2	-9	2	-3	-9

11	12	13	14	15	16	17	18	19	20
9	4	8	6	5	3	7	12	1	8
10	10	11	1	10	1	2	2	13	-3
-5	-9	-14	-2	9	10	-4	-9	-9	10
-9	3	9	9	-12	-9	9	3	4	9

21	22	23	24	25	26	27	28	29	30
7	9	2	11	12	10	15	8	1	2
-2	-4	12	3	1	2	9	1	6	2
9	9	-9	-9	1	2	-4	-4	-2	10
-1	-3	2	1	-9	-9	3	9	9	-9

Exercises: Mixed Friend 9 and Big Friends + Little Friends

1	2	3	4	5	6	7	8	9	10
2	6	1	5	8	3	9	4	7	6
3	4	13	9	2	3	3	4	3	3
9	4	-9	3	4	-1	3	-3	4	-4
1	-9	5	-4	-9	9	-1	9	-9	9
-6	-1	-7	-9	3	1	-9	2	-2	6

11	12	13	14	15	16	17	18	19	20
8	5	4	2	7	3	6	1	9	15
9	2	1	8	8	8	5	3	-4	-1
-3	8	9	4	9	3	-7	10	9	-9
-9	9	1	-9	-4	-9	1	-9	1	3
4	-5	-2	-1	-8	-1	9	3	-7	2

21	22	23	24	25	26	27	28	29	30
3	7	6	5	9	2	1	4	8	3
7	-3	-1	-1	-7	4	8	3	7	8
-5	10	9	4	3	4	6	11	9	3
9	-9	3	-3	9	4	9	-4	-12	-9
3	2	3	9	2	-9	4	-9	3	-2

Exercises: Mixed Friends 6, 7, 8, & 9

1	2	3	4	5	6	7	8	9	10
1	2	6	7	4	9	5	3	4	3
3	2	1	7	2	8	1	8	3	10
2	1	7	3	6	-5	7	2	7	-7
7	7	1	6	-7	-7	-6	-7	-6	6
2	-6	6	2	6	6	3	6	2	-7

11	12	13	14	15	16	17	18	19	20
7	8	2	5	9	3	5	1	3	6
-2	4	3	-2	4	3	8	1	3	6
7	-7	7	9	-7	7	1	3	8	-7
-6	6	-6	-7	6	-6	-7	8	-6	8
7	-4	4	6	4	7	6	-6	4	-5

21	22	23	24	25	26	27	28	29	30
7	3	9	5	6	1	4	2	5	3
7	2	4	-1	7	5	1	7	1	4
-8	6	-8	9	-8	8	8	4	8	6
7	2	7	-8	-3	-6	-7	-8	-7	-8
-8	-8	-8	7	9	7	6	6	6	7

Exercises: Mixed Friends 6, 7, 8, & 9

1	2	3	4	5	6	7	8	9	10
5	1	7	2	3	9	4	3	2	6
4	3	6	6	2	3	9	1	3	8
4	9	-8	6	8	3	-8	9	9	-9
-8	-8	6	-8	-7	8	7	-8	-7	7
7	7	-8	4	8	-6	-6	7	6	-9

11	12	13	14	15	16	17	18	19	20
1	5	8	3	9	9	4	7	6	8
13	9	6	3	3	3	4	7	3	-1
-9	-6	-9	8	3	3	6	-9	-4	7
6	3	8	-9	9	7	-9	-2	9	-9
-7	-6	-6	8	-8	-6	2	8	-6	4

21	22	23	24	25	26	27	28	29	30
5	4	2	7	3	9	5	6	2	1
2	1	8	8	8	-4	8	8	4	8
8	9	4	9	3	9	2	-9	8	6
9	-8	-9	-6	-9	8	9	-3	-9	9
-7	7	6	8	7	-7	1	8	8	-7

Exercises: Big Friends, Little Friends and Mix Friends

1	2	3	4	5	6	7	8	9	10
2	5	9	1	4	7	3	8	6	5
8	7	5	9	3	4	2	6	7	8
-5	-6	-7	3	6	-6	9	-9	4	2
7	-2	-3	-8	3	8	-8	-1	7	9
-9	7	2	-4	7	-6	4	6	-9	-6

11	12	13	3	15	16	17	18	19	20
8	4	6	3	7	5	2	1	9	3
3	4	8	4	6	9	4	4	4	3
-9	8	-9	7	-8	3	8	7	-7	7
3	-7	6	1	9	6	-9	-6	8	-8
8	8	4	-2	1	-8	-1	8	4	-1

21	22	23	24	25	26	27	28	29	30
3	1	9	5	2	6	8	4	7	2
8	3	3	6	5	6	4	9	5	9
-6	4	-7	-3	7	-7	-6	-8	-6	-6
8	6	8	-4	-9	9	7	9	8	9
3	-9	2	6	7	1	2	-6	1	3

Exercises: Big Friends, Little Friends and Mix Friends

1	2	3	4	5	6	7	8	9	10
4	7	2	5	3	8	1	6	5	9
2	4	6	8	3	6	3	4	6	-1
6	-8	6	-7	6	-9	3	-2	4	5
-7	9	-8	4	7	8	7	6	-1	-7
6	-6	-2	-9	8	-7	-9	1	7	8

11	12	13	14	15	16	17	18	19	20
9	6	1	4	2	5	7	3	8	4
7	6	2	4	5	6	6	2	3	1
-3	-7	7	9	7	-3	-5	7	-4	9
4	8	-6	-3	8	-4	6	8	-3	2
7	2	7	1	-6	6	3	-1	9	-9

21	22	23	24	25	26	27	28	29	30
3	5	8	9	6	4	3	1	7	2
4	7	7	2	7	3	4	5	5	4
9	4	-1	-6	4	7	8	4	-6	4
-6	-6	-9	8	-8	-9	-6	-6	7	-3
-4	-9	7	2	7	8	2	8	-8	7

Exercises: Big Friends, Little Friends and Mix Friends

1	2	3	4	5	6	7	8	9	10
5	1	6	3	8	4	2	7	9	3
4	8	9	9	6	9	4	5	2	7
6	3	-8	-8	-9	-6	4	-3	-3	5
-8	3	7	1	6	7	-5	4	6	7
7	-7	-8	6	-7	1	6	-8	2	-6

11	12	13	14	15	16	17	18	19	20
3	7	4	9	2	1	5	8	6	2
4	4	8	-7	6	1	-1	3	7	1
7	-6	4	8	8	3	8	-2	3	8
-9	8	-6	3	-3	5	3	7	4	-7
7	4	7	5	-7	-2	-4	-4	-8	1

21	22	23	24	25	26	27	28	29	30
6	1	5	8	3	7	2	4	9	1
4	6	9	4	2	7	3	6	-3	7
-2	6	-7	3	8	-8	7	-5	4	-4
5	-8	4	-7	-7	9	-6	9	-7	7
-7	6	-7	8	-1	-1	5	1	9	-6

Exercises: Big Friends, Little Friends and Mix Friends

1	2	3	4	5	6	7	8	9	10
7	9	3	6	2	8	5	1	4	8
5	7	8	8	3	2	9	6	9	3
2	-8	-6	2	6	-1	3	9	-8	-5
6	3	-4	-7	-9	7	-9	-3	7	4
-9	4	9	6	8	-4	6	-8	3	-6

11	12	13	14	15	16	17	18	19	20
4	8	2	5	7	3	9	6	1	7
4	5	8	3	7	4	4	7	7	8
6	7	-3	7	-8	8	3	-8	-4	-9
8	-6	6	-6	4	-6	4	7	1	1
-7	-9	-7	4	-9	9	-6	3	-3	4

21	22	23	24	25	26	27	28	29	30
6	1	9	4	8	2	5	3	7	5
5	4	5	8	6	4	8	2	6	7
-8	3	-6	-7	3	5	3	9	-8	-6
7	2	4	-4	-4	-9	4	-8	7	4
-2	-8	-3	9	2	8	-9	-1	3	-2

Yeah.......

You MADE it!

Let's move on to

How to Abacus 3 :

: All about Times and Division

Answer Key

P. 12

1	2	3	4	5	6	7	8
10	**20**	**30**	**40**	**60**	**70**	**80**	**90**

1	2	3	4	5	6	7	8	9	10
10	**10**	**10**	**10**	**10**	**10**	**10**	**10**	**20**	**20**

p.13

1	2	3	4	5	6	7	8
9	**19**	**29**	**39**	**59**	**69**	**79**	**89**

1	2	3	4	5	6	7	8	9	10
9	**9**	**9**	**9**	**19**	**9**	**9**	**9**	**9**	**9**

p.14

1	2	3	4	5	6	7	8
10	**11**	**12**	**13**	**15**	**16**	**17**	**18**

1	2	3	4	5	6	7	8	9	10
12	**10**	**11**	**11**	**15**	**16**	**11**	**10**	**16**	**17**

p.15

1	2	3	4	5	6	7	8
1	**2**	**3**	**4**	**7**	**8**	**9**	**11**

1	2	3	4	5	6	7	8	9	10
2	**1**	**2**	**3**	**3**	**6**	**7**	**6**	**4**	**3**

p.16

1	2	3	4	5	6	7	8	9	10
17	**15**	**7**	**16**	**19**	**9**	**7**	**4**	**6**	**1**
11	12	13	14	15	16	17	18	19	20
1	**7**	**16**	**9**	**10**	**10**	**18**	**9**	**4**	**18**
21	22	23	24	25	26	27	28	29	30
9	**19**	**9**	**18**	**17**	**20**	**14**	**9**	**19**	**12**

p.17

1	2	3	4	5	6	7	8	9	10
9	**25**	**9**	**19**	**15**	**18**	**1**	**1**	**16**	**1**
11	12	13	14	15	16	17	18	19	20
19	**9**	**18**	**9**	**1**	**2**	**14**	**26**	**1**	**23**
21	22	23	24	25	26	27	28	29	30
20	**9**	**0**	**18**	**9**	**26**	**11**	**0**	**18**	**19**

p.18

1	2	3	4	5	6	7	8
10	**20**	**30**	**40**	**11**	**21**	**31**	**41**

1	2	3	4	5	6	7	8	9	10
10	**10**	**10**	**10**	**11**	**11**	**11**	**20**	**21**	**20**

p.19

1	2	3	4	5	6	7	8
8	**18**	**28**	**38**	**9**	**19**	**29**	**39**

1	2	3	4	5	6	7	8	9	10
9	**8**	**9**	**9**	**9**	**9**	**9**	**8**	**8**	**19**

p.20

1	2	3	4	5	6	7	8
10	**11**	**12**	**15**	**16**	**17**	**20**	**25**

1	2	3	4	5	6	7	8	9	10
11	**12**	**10**	**10**	**17**	**15**	**10**	**11**	**21**	**20**

p.21

1	2	3	4	5	6	7	8
2	**3**	**4**	**9**	**12**	**13**	**14**	**19**

1	2	3	4	5	6	7	8	9	10
4	**2**	**3**	**4**	**4**	**7**	**8**	**7**	**3**	**4**

p.22

1	2	3	4	5	6	7	8	9	10
20	**17**	**4**	**12**	**18**	**3**	**16**	**17**	**19**	**25**
11	12	13	14	15	16	17	18	19	20
15	**17**	**18**	**15**	**16**	**10**	**9**	**10**	**10**	**8**
21	22	23	24	25	26	27	28	29	30
15	**8**	**15**	**2**	**10**	**8**	**9**	**10**	**11**	**18**

p.23

1	2	3	4	5	6	7	8	9	10
8	**6**	**16**	**8**	**7**	**9**	**4**	**20**	**21**	**20**
11	12	13	14	15	16	17	18	19	20
9	**9**	**8**	**4**	**9**	**2**	**16**	**8**	**7**	**17**
21	22	23	24	25	26	27	28	29	30
18	**4**	**26**	**4**	**15**	**9**	**20**	**20**	**17**	**6**

p.24

1	2	3	4	5	6	7	8	9	10
20	**8**	**8**	**4**	**10**	**16**	**8**	**10**	**12**	**17**
11	12	13	14	15	16	17	18	19	20
4	**1**	**7**	**7**	**8**	**9**	**9**	**19**	**2**	**12**
21	22	23	24	25	26	27	28	29	30
2	**20**	**8**	**2**	**9**	**8**	**10**	**9**	**7**	**6**

p.25

1	2	3	4	5	6	7	8	9	10
27	**15**	**20**	**10**	**13**	**16**	**18**	**1**	**9**	**8**
11	12	13	14	15	16	17	18	19	20
10	**10**	**2**	**9**	**2**	**6**	**4**	**11**	**9**	**16**
21	22	23	24	25	26	27	28	29	30
3	**10**	**12**	**11**	**15**	**10**	**18**	**17**	**10**	**10**

p.26

1	2	3	4	5	6	7	8		
10	**11**	**12**	**20**	**21**	**22**	**30**	**31**		
1	2	3	4	5	6	7	8	9	10
10	**11**	**12**	**10**	**11**	**10**	**12**	**12**	**11**	**12**

p.27

1	2	3	4	5	6	7	8		
7	**8**	**9**	**17**	**18**	**19**	**27**	**28**		
1	2	3	4	5	6	7	8	9	10
9	**7**	**8**	**9**	**9**	**8**	**9**	**9**	**7**	**9**

p.28

1	2	3	4	5	6	7	8		
10	**15**	**16**	**20**	**25**	**26**	**30**	**35**		
1	2	3	4	5	6	7	8	9	10
15	**16**	**16**	**10**	**11**	**11**	**15**	**15**	**16**	**16**

p.29

1	2	3	4	5	6	7	8		
3	**4**	**8**	**9**	**13**	**14**	**18**	**19**		
1	2	3	4	5	6	7	8	9	10
4	**3**	**4**	**4**	**4**	**4**	**3**	**8**	**9**	**8**

p.30

1	2	3	4	5	6	7	8	9	10
20	**20**	**7**	**3**	**4**	**4**	**9**	**7**	**20**	**8**
11	12	13	14	15	16	17	18	19	20
8	**3**	**3**	**13**	**10**	**9**	**10**	**7**	**14**	**16**
21	22	23	24	25	26	27	28	29	30
13	**14**	**18**	**7**	**10**	**25**	**12**	**3**	**4**	**11**

p.31

1	2	3	4	5	6	7	8	9	10
9	**7**	**9**	**16**	**4**	**8**	**17**	**8**	**26**	**15**
11	12	13	14	15	16	17	18	19	20
8	**4**	**21**	**7**	**9**	**15**	**25**	**8**	**15**	**15**
21	22	23	24	25	26	27	28	29	30
4	**6**	**10**	**4**	**7**	**11**	**8**	**8**	**10**	**15**

p.32

1	2	3	4	5	6	7	8	9	10
13	**8**	**11**	**9**	**7**	**3**	**10**	**8**	**10**	**4**
11	12	13	14	15	16	17	18	19	20
12	**9**	**4**	**7**	**7**	**11**	**10**	**6**	**7**	**17**
21	22	23	24	25	26	27	28	29	30
3	**4**	**7**	**12**	**1**	**9**	**3**	**8**	**3**	**2**

p.33

1	2	3	4	5	6	7	8	9	10
15	**11**	**13**	**8**	**5**	**16**	**8**	**17**	**17**	**3**
11	12	13	14	15	16	17	18	19	20
7	**10**	**7**	**16**	**8**	**26**	**3**	**9**	**16**	**17**
21	22	23	24	25	26	27	28	29	30
18	**16**	**16**	**12**	**5**	**9**	**26**	**16**	**12**	**15**

p.34

1	2	3	4	5	6	7	8		
10	**11**	**12**	**13**	**20**	**21**	**22**	**23**		
1	2	3	4	5	6	7	8	9	10
11	**13**	**12**	**13**	**12**	**12**	**13**	**13**	**11**	**10**

p.35

1	2	3	4	5	6	7	8		
6	**7**	**8**	**9**	**16**	**17**	**18**	**19**		
1	2	3	4	5	6	7	8	9	10
7	**6**	**7**	**7**	**6**	**7**	**8**	**8**	**8**	**9**

p.36

1	2	3	4	5	6	7	8		
10	**15**	**20**	**25**	**30**	**35**	**40**	**45**		
1	2	3	4	5	6	7	8	9	10
15	**15**	**15**	**15**	**10**	**10**	**10**	**15**	**15**	**15**

p.37

1	2	3	4	5	6	7	8		
4	**9**	**14**	**19**	**24**	**29**	**34**	**39**		
1	2	3	4	5	6	7	8	9	10
9	**9**	**9**	**4**	**4**	**4**	**4**	**9**	**4**	**4**

p.38

1	2	3	4	5	6	7	8	9	10
7	**6**	**4**	**6**	**4**	**4**	**10**	**4**	**14**	**15**
11	12	13	14	15	16	17	18	19	20
18	**11**	**19**	**4**	**19**	**21**	**18**	**17**	**17**	**6**
21	22	23	24	25	26	27	28	29	30
14	**14**	**10**	**10**	**6**	**4**	**17**	**16**	**12**	**10**

p.39

1	2	3	4	5	6	7	8	9	10
26	**24**	**4**	**22**	**9**	**6**	**21**	**22**	**4**	**17**
11	12	13	14	15	16	17	18	19	20
4	**9**	**22**	**4**	**9**	**4**	**21**	**22**	**7**	**23**
21	22	23	24	25	26	27	28	29	30
26	**18**	**9**	**7**	**9**	**4**	**2**	**4**	**6**	**22**

p.40

1	2	3	4	5	6	7	8	9	10
15	**12**	**9**	**15**	**10**	**8**	**10**	**15**	**6**	**15**
11	12	13	14	15	16	17	18	19	20
15	**4**	**16**	**19**	**6**	**9**	**6**	**7**	**17**	**6**
21	22	23	24	25	26	27	28	29	30
16	**4**	**16**	**10**	**16**	**6**	**4**	**20**	**10**	**6**

p.41

1	2	3	4	5	6	7	8	9	10
9	**6**	**2**	**13**	**6**	**20**	**2**	**25**	**15**	**6**
11	12	13	14	15	16	17	18	19	20
22	**16**	**7**	**11**	**6**	**14**	**11**	**4**	**9**	**11**
21	22	23	24	25	26	27	28	29	30
3	**16**	**10**	**22**	**13**	**14**	**15**	**2**	**4**	**21**

p.43

1	2	3	4	5	6	7	8	9	10
3	**10**	**10**	**5**	**9**	**14**	**7**	**16**	**8**	**8**
11	12	13	14	15	16	17	18	19	20
6	**15**	**14**	**11**	**15**	**4**	**7**	**9**	**8**	**7**
21	22	23	24	25	26	27	28	29	30
11	**8**	**5**	**4**	**22**	**16**	**10**	**19**	**6**	**9**

p.44

1	2	3	4	5	6	7	8		
5	**15**	**25**	**35**	**45**	**55**	**65**	**75**		
1	2	3	4	5	6	7	8	9	10
5	**5**	**5**	**15**	**15**	**15**	**15**	**15**	**25**	**25**

p.45

1	2	3	4	5	6	7	8		
4	**14**	**24**	**34**	**44**	**54**	**64**	**74**		
1	2	3	4	5	6	7	8	9	10
4	**4**	**4**	**4**	**14**	**18**	**14**	**14**	**24**	**24**

p.46

1	2	3	4	5	6	7	8		
5	**6**	**7**	**8**	**15**	**16**	**17**	**18**		
1	2	3	4	5	6	7	8	9	10
8	**8**	**6**	**5**	**6**	**7**	**8**	**17**	**15**	**16**

p.47

1	2	3	4	5	6	7	8		
1	**2**	**3**	**4**	**11**	**12**	**13**	**14**		
1	2	3	4	5	6	7	8	9	10
1	**1**	**2**	**3**	**1**	**1**	**2**	**4**	**2**	**3**

p.48

1	2	3	4	5	6	7	8	9	10
9	**7**	**4**	**8**	**4**	**5**	**1**	**5**	**1**	**3**
11	12	13	14	15	16	17	18	19	20
5	**5**	**5**	**5**	**3**	**4**	**1**	**7**	**4**	**3**
21	22	23	24	25	26	27	28	29	30
2	**4**	**8**	**4**	**3**	**19**	**4**	**9**	**9**	**7**

p.49

1	2	3	4	5	6	7	8	9	10
5	**4**	**14**	**5**	**10**	**9**	**5**	**10**	**16**	**20**
11	12	13	14	15	16	17	18	19	20
9	**13**	**16**	**8**	**16**	**11**	**10**	**20**	**18**	**10**
21	22	23	24	25	26	27	28	29	30
15	**20**	**9**	**1**	**6**	**4**	**14**	**2**	**5**	**14**

p.50

1	2	3	4	5	6	7	8		
5	**6**	**15**	**16**	**25**	**26**	**35**	**36**		
1	2	3	4	5	6	7	8	9	10
5	**5**	**5**	**5**	**5**	**6**	**6**	**6**	**6**	**15**

p.51

1	2	3	4	5	6	7	8		
3	**4**	**13**	**14**	**23**	**24**	**33**	**34**		
1	2	3	4	5	6	7	8	9	10
3	**3**	**3**	**3**	**4**	**2**	**4**	**4**	**4**	**4**

p.52

1	2	3	4	5	6	7	8		
5	**6**	**7**	**15**	**16**	**17**	**25**	**26**		
1	2	3	4	5	6	7	8	9	10
5	**5**	**5**	**6**	**5**	**7**	**6**	**6**	**6**	**7**

p.53

1	2	3	4	5	6	7	8		
2	**3**	**4**	**12**	**13**	**14**	**22**	**23**		
1	2	3	4	5	6	7	8	9	10
2	**2**	**3**	**2**	**3**	**2**	**4**	**3**	**4**	**0**

p.54

1	2	3	4	5	6	7	8	9	10
4	**5**	**5**	**4**	**3**	**3**	**4**	**2**	**3**	**3**
11	12	13	14	15	16	17	18	19	20
3	**4**	**9**	**2**	**7**	**4**	**2**	**6**	**3**	**16**
21	22	23	24	25	26	27	28	29	30
3	**3**	**6**	**3**	**8**	**4**	**6**	**3**	**6**	**8**

p.55

1	2	3	4	5	6	7	8	9	10
11	**12**	**22**	**6**	**10**	**9**	**9**	**5**	**5**	**5**
11	12	13	14	15	16	17	18	19	20
3	**2**	**15**	**5**	**13**	**1**	**2**	**2**	**4**	**4**
21	22	23	24	25	26	27	28	29	30
9	**4**	**3**	**2**	**5**	**3**	**13**	**15**	**14**	**20**

p.56

1	2	3	4	5	6	7	8	9	10
2	**5**	**11**	**8**	**5**	**5**	**4**	**10**	**10**	**6**
11	12	13	14	15	16	17	18	19	20
3	**10**	**12**	**15**	**20**	**3**	**5**	**11**	**20**	**10**
21	22	23	24	25	26	27	28	29	30
4	**1**	**10**	**4**	**5**	**4**	**1**	**13**	**20**	**3**

p.57

1	2	3	4	5	6	7	8	9	10
4	**5**	**8**	**10**	**2**	**3**	**2**	**3**	**5**	**2**
11	12	13	14	15	16	17	18	19	20
4	**4**	**15**	**14**	**4**	**11**	**4**	**4**	**10**	**9**
21	22	23	24	25	26	27	28	29	30
10	**10**	**5**	**5**	**4**	**2**	**2**	**15**	**16**	**6**

p.58

1	2	3	4	5	6	7	8		
11	**12**	**13**	**14**	**21**	**22**	**23**	**24**		
1	2	3	4	5	6	7	8	9	10
12	**11**	**12**	**14**	**12**	**13**	**14**	**12**	**14**	**13**

p.59

1	2	3	4	5	6	7	8		
5	**6**	**7**	**8**	**15**	**16**	**17**	**18**		
1	2	3	4	5	6	7	8	9	10
5	**6**	**8**	**6**	**7**	**7**	**7**	**5**	**5**	**8**

p.60

1	2	3	4	5	6	7	8	9	10
23	**10**	**10**	**10**	**18**	**9**	**11**	**14**	**13**	**5**
11	12	13	14	15	16	17	18	19	20
13	**14**	**29**	**14**	**7**	**18**	**7**	**13**	**24**	**19**
21	22	23	24	25	26	27	28	29	30
18	**13**	**8**	**6**	**8**	**14**	**9**	**8**	**13**	**8**

p.61

1	2	3	4	5	6	7	8	9	10
12	**10**	**24**	**10**	**8**	**11**	**9**	**10**	**15**	**16**
11	12	13	14	15	16	17	18	19	20
22	**10**	**16**	**8**	**21**	**6**	**3**	**24**	**2**	**22**
21	22	23	24	25	26	27	28	29	30
14	**31**	**14**	**6**	**20**	**21**	**6**	**21**	**7**	**15**

p.62

1	2	3	4	5	6	7	8		
12	**13**	**14**	**22**	**23**	**24**	**32**	**33**		
1	2	3	4	5	6	7	8	9	10
13	**12**	**12**	**12**	**13**	**14**	**13**	**13**	**13**	**14**

p.63

1	2	3	4	5	6	7	8		
5	**6**	**7**	**15**	**16**	**17**	**25**	**26**		
1	2	3	4	5	6	7	8	9	10
5	**5**	**7**	**5**	**6**	**6**	**6**	**5**	**6**	**7**

p.64

1	2	3	4	5	6	7	8	9	10
24	**9**	**10**	**10**	**13**	**8**	**11**	**14**	**12**	**2**
11	12	13	14	15	16	17	18	19	20
14	**5**	**6**	**8**	**12**	**14**	**10**	**14**	**13**	**3**
21	22	23	24	25	26	27	28	29	30
12	**8**	**13**	**13**	**6**	**13**	**10**	**8**	**8**	**6**

p.65

1	2	3	4	5	6	7	8	9	10
10	**15**	**17**	**20**	**11**	**15**	**11**	**1**	**9**	**13**
11	12	13	14	15	16	17	18	19	20
2	**5**	**24**	**16**	**13**	**9**	**10**	**4**	**3**	**9**
21	22	23	24	25	26	27	28	29	30
16	**15**	**4**	**11**	**16**	**4**	**4**	**9**	**4**	**11**

p.66

1	2	3	4	5	6	7	8		
13	**14**	**23**	**24**	**33**	**34**	**43**	**44**		
1	2	3	4	5	6	7	8	9	10
14	**13**	**13**	**14**	**23**	**24**	**23**	**13**	**14**	**14**

p.67

1	2	3	4	5	6	7	8		
5	**6**	**15**	**16**	**25**	**26**	**35**	**36**		
1	2	3	4	5	6	7	8	9	10
5	**6**	**6**	**6**	**5**	**5**	**5**	**6**	**15**	**16**

p.68

1	2	3	4	5	6	7	8	9	10
10	**12**	**6**	**19**	**7**	**14**	**12**	**1**	**10**	**12**
11	12	13	14	15	16	17	18	19	20
10	**14**	**12**	**11**	**5**	**6**	**14**	**5**	**6**	**18**
21	22	23	24	25	26	27	28	29	30
14	**13**	**6**	**7**	**6**	**14**	**5**	**20**	**7**	**10**

p.69

1	2	3	4	5	6	7	8	9	10
19	**15**	**16**	**7**	**11**	**13**	**8**	**2**	**18**	**2**
11	12	13	14	15	16	17	18	19	20
5	**7**	**4**	**10**	**8**	**15**	**4**	**13**	**7**	**10**
21	22	23	24	25	26	27	28	29	30
4	**23**	**2**	**9**	**22**	**23**	**0**	**10**	**5**	**8**

p.70

1	2	3	4	5	6	7	8		
14	**24**	**34**	**44**	**54**	**64**	**74**	**84**		
1	2	3	4	5	6	7	8	9	10
24	**24**	**34**	**34**	**44**	**44**	**54**	**54**	**64**	**64**

p.71

1	2	3	4	5	6	7	8		
5	**15**	**25**	**35**	**45**	**55**	**65**	**75**		
1	2	3	4	5	6	7	8	9	10
5	**5**	**5**	**5**	**15**	**15**	**15**	**15**	**25**	**25**

p.72

1	2	3	4	5	6	7	8	9	10
13	**6**	**5**	**20**	**5**	**12**	**5**	**7**	**11**	**5**
11	12	13	14	15	16	17	18	19	20
5	**8**	**14**	**14**	**12**	**5**	**14**	**8**	**9**	**24**
21	22	23	24	25	26	27	28	29	30
13	**11**	**7**	**6**	**5**	**4**	**23**	**14**	**14**	**5**

p.73

1	2	3	4	5	6	7	8	9	10
9	**4**	**3**	**4**	**8**	**15**	**5**	**16**	**3**	**20**
11	12	13	14	15	16	17	18	19	20
9	**19**	**13**	**4**	**12**	**4**	**14**	**8**	**8**	**10**
21	22	23	24	25	26	27	28	29	30
17	**7**	**20**	**14**	**16**	**5**	**28**	**5**	**15**	**3**

p.74

1	2	3	4	5	6	7	8	9	10
15	**6**	**21**	**25**	**11**	**11**	**10**	**12**	**10**	**5**
11	12	13	14	15	16	17	18	19	20
13	**7**	**10**	**11**	**16**	**14**	**13**	**7**	**12**	**8**
21	22	23	24	25	26	27	28	29	30
5	**5**	**4**	**12**	**11**	**15**	**12**	**11**	**13**	**12**

p.75

1	2	3	4	5	6	7	8	9	10
12	**12**	**3**	**10**	**14**	**17**	**6**	**12**	**13**	**3**
11	12	13	14	15	16	17	18	19	20
4	**5**	**7**	**13**	**16**	**16**	**7**	**11**	**8**	**9**
21	22	23	24	25	26	27	28	29	30
17	**13**	**11**	**26**	**12**	**15**	**25**	**10**	**13**	**17**

p.76

1	2	3	4	5	6	7	8	9	10
3	**11**	**6**	**1**	**23**	**7**	**10**	**10**	**15**	**18**
11	12	13	3	15	16	17	18	19	20
13	**17**	**15**	**13**	**15**	**15**	**4**	**14**	**18**	**4**
21	22	23	24	25	26	27	28	29	30
16	**5**	**15**	**10**	**12**	**15**	**15**	**8**	**15**	**17**

p.77

1	2	3	4	5	6	7	8	9	10
11	**6**	**4**	**1**	**27**	**6**	**5**	**15**	**21**	**14**
11	12	13	14	15	16	17	18	19	20
24	**15**	**11**	**15**	**16**	**10**	**17**	**19**	**13**	**7**
21	22	23	24	25	26	27	28	29	30
6	**1**	**12**	**15**	**16**	**13**	**11**	**12**	**5**	**14**

p.78

1	2	3	4	5	6	7	8	9	10
14	**8**	**6**	**11**	**4**	**15**	**11**	**5**	**16**	**16**
11	12	13	14	15	16	17	18	19	20
12	**17**	**17**	**18**	**6**	**8**	**11**	**12**	**12**	**5**
21	22	23	24	25	26	27	28	29	30
6	**11**	**4**	**16**	**5**	**14**	**11**	**15**	**12**	**5**

p.79

1	2	3	4	5	6	7	8	9	10
11	**15**	**10**	**15**	**10**	**12**	**14**	**5**	**15**	**4**
11	12	13	14	15	16	17	18	19	20
15	**5**	**6**	**13**	**1**	**18**	**14**	**15**	**2**	**11**
21	22	23	24	25	26	27	28	29	30
8	**2**	**9**	**10**	**15**	**10**	**11**	**5**	**15**	**8**

Formula Chart

Big Friends

+ 1 = - 9 + 10	- 1 = - 10 + 9
+ 2 = - 8 + 10	- 2 = - 10 + 8
+ 3 = - 7 + 10	- 3 = - 10 + 7
+ 4 = - 6 + 10	- 4 = - 10 + 6
+ 5 = - 5 + 10	- 5 = - 10 + 5
+ 6 = - 4 + 10	- 6 = - 10 + 4
+ 7 = - 3 + 10	- 7 = - 10 + 3
+ 8 = - 2 + 10	- 8 = - 10 + 2
+ 9 = - 1 + 10	- 9 = - 10 + 1

Little Friends

+ 1 = + 5 – 4	- 1 = - 5 + 4
+ 2 = + 5 – 3	- 2 = - 5 + 3
+ 3 = + 5 – 2	- 3 = - 5 + 2
+ 4 = + 5 – 1	- 4 = - 5 + 1

Mixed Friends

+ 6 = - 5 + 1 + 10	- 6 = - 10 + 5 -1
+ 7 = - 5 + 2 + 10	- 7 = - 10 + 5 - 2
+ 8 = - 5 + 3 + 10	- 8 = - 10 + 5 - 3
+ 9 = - 5 + 4 + 10	- 9 = - 10 + 5 - 4

Made in the USA
Las Vegas, NV
11 January 2026

39111680R00048